FEBRUARY

M	T	W	T	F	S	S
		1	2	3	4	5
6	7	8	9	10	11	12
13	14	15	16	17	18	19
20	21	22	23	24	25	26
27	28					

APRIL

M	T	W	T	F	S	S
					1	2
3	4	5	6	7	8	9
10	11	12	13	14	15	16
17	18	19	20	21	22	23
24	25	26	27	28	29	30

JUNE

M	T	W	T	F	S	S
			1	2	3	4
5	6	7	8	9	10	11
12	13	14	15	16	17	18
19	20	21	22	23	24	25
26	27	28	29	30		

THE
ALMANAC

THE
ALMANAC

A SEASONAL GUIDE TO
2023

LIA LEENDERTZ

With illustrations by Whooli Chen

Gaia

First published in Great Britain in 2022 by Gaia, an imprint of
Octopus Publishing Group Ltd
Carmelite House, 50 Victoria Embankment, London EC4Y 0DZ
www.octopusbooks.co.uk

An Hachette UK Company
www.hachette.co.uk

ISBN 978-1-85675-463-7

A CIP catalogue record for this book is available from the British Library.

Printed and bound in the United Kingdom.

10 9 8 7 6 5 4 3 2 1

Publisher: Stephanie Jackson
Creative Director: Jonathan Christie
Designer: Matt Cox at Newman+Eastwood
Editor: Sarah Kyle
Copy Editor: Alison Wormleighton
Senior Production Manager: Peter Hunt

This FSC® label means that materials used for the product have been
responsibly sourced

Ovens should be preheated to the specific temperature – if using a fan-assisted oven,
follow manufacturer's instructions for adjusting the time and the temperature.
Pepper should be freshly ground black pepper unless otherwise stated.

CONTENTS

INTRODUCTION

Welcome to *The Almanac: A Seasonal Guide to 2023*. Hello to new readers – I hope you will find something within this book to help you enjoy every month of the year ahead. And if you are a regular reader, then thank you for keeping me company through another year.

For this year's theme I have cast my eyes up to the heavens, and to the invisible and occasionally visible patterns of the solar system and zodiac that churn around us throughout the year. We will look at the planets of our solar system, what and where they are in relation to earth, and when is the best time to spot each in our skies. We also visit the grand mythology behind the zodiac, and match each zodiacal creature to some very earthy and bawdy British folk songs. Watch out for Cancer's crab…

This almanac is as much about minutiae as it is about grandeur, and we will also follow the swirling micro world of the garden pond through the year. Mating frogs, maturing tadpoles, emerging dragonflies, flowering water mint, swooping pipistrelle bats and falling leaves – the garden pond acts as a microcosm of the seasonal changes we see in the wider landscape, and I hope you will enjoy a monthly peer into its depths.

In addition to this you will find many ways of connecting to the season and the months: moon phase charts so that you will know the exact moment of the full Plough Moon or Rose Moon; sunrise and set tables so that you can follow the tip of the northern hemisphere towards and then away from the sun; jobs to carry out in the garden; tide timetables; spring and neap tides; and seasonal and celebration recipes – including this year a 'bun of the month', from Swedish *semlor* for Fat Tuesday to hot cross buns for Easter and sweet, round *challah* for Rosh Hashana.

Whether you are a baker, a gardener, a singer, a mudlark, a stargazer, a nature lover or just someone who gets an odd thrill out of looking at tide timetables (that's me), I hope that within this almanac you will find your own keys to unlocking some special moments in the year ahead. Have a wonderful 2023.

Lia Leendertz

THE YEAR AHEAD

The year

Gregorian year	2023, begins 1st January
Japanese year	2683, begins 1st January
Chinese year	Black Water Rabbit, begins 22nd January
Islamic year	1445, begins 18th July
Coptic year	1740, begins 12th September
Jewish year	5784, begins 15th September

The sky at night in the year ahead

It should be a good year for spotting the bright planets. Venus starts the year as an evening star, visible for a short time after sunset. It will then become steadily more prominent, setting some four hours after sunset by mid-May. From then on it will approach the sun again and be lost in the glare of our sunset by mid-July. It will reappear as a morning star by the end of August, getting steadily higher until it rises some four hours before sunrise by the end of October.

Mars will be bright and high at the start of the year but fading as the weeks go by. It will be dim by mid-March, then unremarkable for the rest of the year.

Jupiter progresses on its 12-year orbit. It will give a good display, reaching a greater height in the night sky than last year. It will get higher each year now until 2025–6, after which it will gradually get lower each year until around 2031 when it will be very low in the sky all year. The cycle will then repeat.

Saturn progresses on its 29-year orbit. It will be dim for the first half of this year but bright by autumn. A small telescope will reveal the ring system, and the time around opposition (see page 172) this year will be the last chance of a good view of the rings for quite a while. They are 'closing' and will be seen edge-on by 2025. They will then 'open' and be visible again by 2026–7.

There will be two solar eclipses this year but sadly neither will be visible from the UK or Ireland. There will also be two lunar eclipses, one of which will be visible from the UK and Ireland, at the end of October. This will only be a partial eclipse, which will be interesting but not awesome.

The full moon will detract from the Quadrantids meteor shower in January, but the Perseids in August and the Geminids in December will have dark skies.

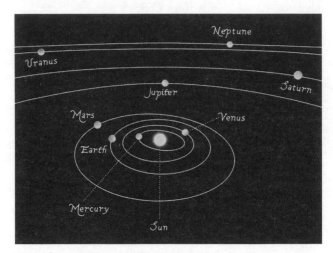

The position of the planets of the solar system on 1st January 2023

Notes on using the tide times

The full tide timetable given each month is for Dover, because Dover is widely used as a standard port from which to work out all other tide times. Every port has a 'high water time difference on Dover' figure, which you can find on the internet. For instance, Bristol's high water time difference on Dover is –4h 10m, and so looking at this almanac's visual tide timetable you would just trace your finger back along it 4 hours and 10 minutes to see that a midday high tide at Dover would mean it will be high tide at Bristol at 07.50. London Bridge's is +2h 52m, so – tracing forwards – a midday high tide at Dover would see a high tide in London at 14.52. Once you know any local port's figure, you can just trace that amount of time backwards or forwards along the Dover tide line.

Here are a few ports and their high water time differences on Dover. Find your local one by searching the name of the port and the phrase 'high water time difference on Dover'.

Aberdeen:	+2h 31m	Cork:	−5h 23m
Firth of Forth:	+3h 50m	Swansea:	−4h 50m
Port Glasgow:	+1h 32m	Bristol:	−4h 10m
Newcastle-upon-Tyne:	+4h 33m	London Bridge:	+2h 52m
Belfast Lough:	+0h 7m	Lyme Regis:	−4h 55m
Hull:	−4h 52m	Newquay:	−6h 4m
Liverpool:	+0h 14m	St Helier, Jersey:	−4h 55m

Do not use these where accuracy is critical; instead, you will need to buy a local tide timetable or subscribe to Easy Tide, www.ukho.gov.uk/easytide. Also note that no timetable will take into account the effects of wind and barometric pressure.

Spring tide and neap tide dates are also included. Spring tides are the most extreme tides of the month – the highest and lowest tides – and neap tides are the least extreme. Spring tides occur as a result of the pull that occurs when the sun, moon and earth are aligned. Alignment occurs at new moon and full moon, but the surge – the spring tide – is slightly delayed because of the mass of water to be moved. It usually follows one to three days after. Knowledge of spring tides is particularly useful if you are a keen rock-pooler, beachcomber or mudlark. You want a low spring tide for best revelations.

General notes
All times in this almanac have been adjusted for British Summer Time/Irish Standard Time, when relevant.

Haltwhistle in Northumberland has been chosen for the sunrise and set tables because it is the midpoint of the longest north–south meridian within the British Isles.

All efforts have been made to ensure the dates within this almanac are correct, but some may be subject to cancellation or rearrangement after publication.

January

1 New Year's Day

2 Bank holiday, England, Wales, Scotland, Northern Ireland, Ireland

3 Bank holiday, Scotland

5 Twelfth Night (Christian/traditional)

6 Epiphany/Three Kings' Day/Little Christmas (Christian)

6 Nollaig na mBan/Women's Christmas (Christian/Irish traditional)

6 Orthodox Christmas Eve (Orthodox)

7 Orthodox Christmas Day (Orthodox)

7 Lidat (Rastafarian)

9 Plough Monday (English traditional)

13 Lohri (Punjabi winter festival)

17 Old Twelfth Night (traditional)

21 Chinese New Year's Eve

22 Chinese New Year/Spring Festival/Lunar New Year – year of the Black Water Rabbit begins

25 Burns Night (Scottish traditional)

25 Vasant Panchami (Hindu spring festival)

28 28th–29th: RSPB Big Garden Birdwatch

JANUARY AT A GLANCE

Spend time outside during the year and you will soon come to see that there is a season for everything, for seeds to tentatively germinate and begin to grow, for frantic work, such as that of the honeybees when the sun shines, for harvesting and feasting, and for closing down, resting and recovering. Our human years, generally, do not cycle. We hit the ground running in early January, let up for a couple of weeks in August and then push on through until the Christmas break. Unlike the birds and the bees, the flowers and the trees, we are forever 'on'.

But the natural year can act as a guide to living, if we let it. And January – it just so happens – is a wonderful place to start, because if you step outside now and look for obvious activity, you will find, well, very little. Yes, there is the odd bumblebee buzzing on the milder days, and the shoots of the snowdrops have pushed through the cold earth, but other than that? There is a whole lot of peace and quiet, of plants and animals tucked away and resting, pulled back in on themselves, surrounded by piles of leaves. In these dark and inhospitable days they conserve their energies, remaining underneath the soil or secreted away in little nooks. It isn't that nothing is happening at all underground: roots stretch down deeper into the soil during all but the hardest frosts. The plants and animals are laying the foundations of the year ahead, but they are doing it from a place of deep slumber.

So this is a time of darkness, rest and recuperation. And the challenge for us busy-busy humans is not to fight it, but to let it be so. Take your opportunities for calm and self-reflection wherever you can find them this month. Close your eyes, breathe and welcome that peace and quiet into your life, even if it is only for ten minutes a day. Cook stews. Stay in and light candles. Sleep more and daydream when you are awake. You are laying the foundations for the year, now. The light will come soon enough. Don't fight the darkness.

THE SKY AT NIGHT

The year begins beautifully with the moon, Mars, Jupiter and Saturn all visible early in the month. Mars was at opposition – that is, at its brightest and closest – in December, and so is still high and bright in the sky. It will be dim by March, so this is the time to catch a glimpse of its fiery hues.

3rd: Close approach of a bright Mars and the Moon. First visible in the east at around 16.20, they reach an altitude of 63 degrees in the south by 21.30 and go on to set in the northwest around 05.00 the next day. Closest at around 20.00 with a separation of less than 1 degree.

3rd–4th: Quadrantids meteor shower. This runs from 1st to 5th January with up to 40 meteors per hour at its peak on the late night of the 4th and early morning of the 5th. Due to the full moon, only the brightest trails will be visible.

22nd: Brief view of close approach of Venus and Saturn. First visible in the dusk at around 17.00 in the southwest at an altitude of 11 degrees with a dim Saturn 0.4 degrees above Venus. They set at around 18.00 in the southwest.

23rd: Close approach of the moon, Venus and Saturn. First visible in the dusk at around 17.00 in the southwest at an altitude of 11 degrees. Visible for one hour before setting.

25th: Close approach of Jupiter and the moon. First visible in the dusk around 17.00 at an altitude of 35 degrees in the south. Visible until setting in the west at around 21.20. Saturn and Venus will also be visible until around 18.00.

26th: Close approach of Mars and the moon. First visible in the southeast at an altitude of 50 degrees at around 17.00. They reach an altitude of 63 degrees in the south by 19.40 and set in the northwest at around 03.50 the next day.

30th: Mercury at greatest western elongation. Mercury, the closest planet to the sun, is often hard to spot when lost in the sun's glare. Towards the end of the month it will be the furthest it gets away from the sun in our sky, so you may spot it low in the eastern sky up to an hour before sunrise.

THE SOLAR SYSTEM

Mars

This is the best moment of this year to see Mars. It was at opposition – at its closest to earth – late last year, and will become dimmer as the year wears on. See it high in the south in the evening hours, early this month, glittering red. The red colour is iron oxide on the surface, like rust. Mars, named after the Roman god of war, is one of the four terrestrial planets (as opposed to the four gaseous planets) and so is more earth-like than most solar system objects, though about half the size of earth.

Mars is one of the five bright planets, often easily visible to the naked eye. Mars is the fourth planet from the sun (earth is the third), so, like earth, it is part of the inner solar system. Not only is it relatively close to us in solar system terms, but it is also not lost in the glare of the sun, as Mercury and Venus often are (being closer to it than to us).

Sunrise and set
Haltwhistle, Northumberland

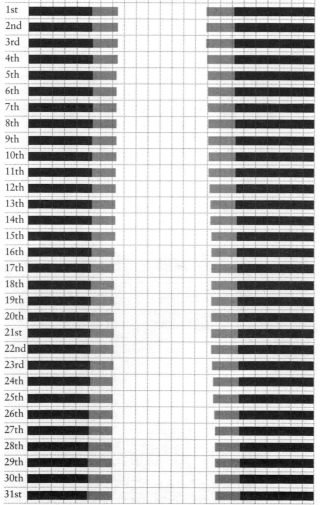

THE SEA

Average sea temperature in Celcius

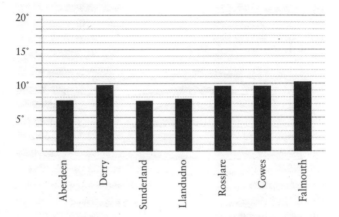

Spring and neap tides

Spring tides are the most extreme tides of the month, with the highest rises and the lowest falls, and they follow a couple of days after the full moon and new moon. These are the times to choose a low tide and go rock-pooling, mudlarking or coastal fossil-hunting. Neap tides are the least extreme, with the smallest movement, and they fall in between the spring tides.

Spring tides: 7th–8th and 23rd–24th

Neap tides: 16th–17th and 29th–30th

Spring tides are shaded in black in the chart opposite.

January tide timetable for Dover

For guidance on how to convert this for your local area, see page 8.

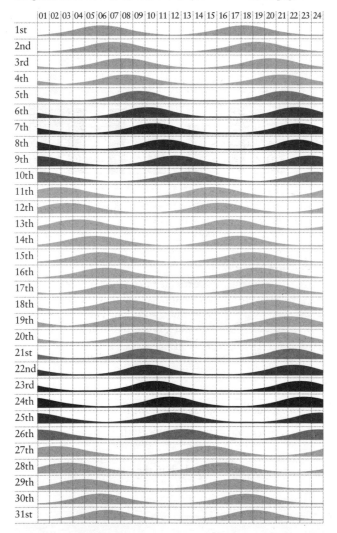

THE MOON

Moon phases

Full moon – 6th January, 23.08

Last quarter – 15th January, 02.10

New moon – 21st January, 20.53

First quarter – 28th January, 15.19

Moonrise and set

Like the sun, the moon rises roughly in the east and sets roughly in the west. It also rises around 50 minutes later each day. Use the following guide to work out approximate moonrise times.

Full moon: Rises near sunset, opposite the sun, so in the east as the sun sets in the west.
Last quarter: Rises around midnight, and is at its highest point as the sun rises.
New moon: Rises at sunrise, in the same part of the sky as the sun (and so cannot be seen).
First quarter: Rises near noon, and is at its highest point as the sun sets.

Full moon

January's full moon is known as the Wolf Moon or the Stay at Home Moon.

New moon

This month's new moon, on the 21st, is in Capricorn. Astrologers believe that the new moon is a quiet, contemplative time before a phase of growth. Each new moon has its own energy, depending on the zodiacal sign that it is in, and the Capricorn new moon is said to rule ambitions and goals.

Moon phases for January

1st	2nd	3rd	4th	5th

6th FULL	7th	8th	9th	10th

11th	12th	13th	14th	15th

16th	17th	18th	19th	20th

21st NEW	22nd	23rd	24th	25th

26th	27th	28th	29th	30th

31st				

GARDENS

To enjoy this month

Ornamental: Hellebores, dogwood stems, willow stems, hazel catkins, seed heads, grasses, sarcococca, witch hazel, crocuses, aconites, holly, ivy and its flowers and berries, box, skeletons of trees, mahonia flowers, winter clematis, cotoneaster and pyracantha berries
Edible: Purple sprouting broccoli, Oriental leaves, Jerusalem artichokes, carrots, leeks, parsnips, swedes, chervil, coriander, parsley, rosemary, sage, bay

Gardening by the moon
The following is a guide to planting with the phases of the moon, according to traditional practices. It also works as a guide to the month's gardening for moon-gardening cynics, who can do these jobs whenever they wish during the month ahead.

First quarter to full moon: 30th December 2022–6th and 29th–5th February
This is the best time for sowing crops that develop above ground, but is bad for root crops. Plant out seedlings and young plants. Take cuttings and make grafts. Avoid any other pruning. Fertilise.
- Sow chillies and aubergines indoors in a heated propagator.
- Sow broad beans straight into the ground if it is not frozen and cover them with cloches.
- Sow hardy peas and sweet peas in pots under cover.

Full moon to last quarter: 7th–14th
A 'drawing down' energy. This phase is a good time for sowing and planting any crops that develop below ground: root crops, bulbs and perennials. Light is high but decreasing.
- Plant garlic and rhubarb crowns if the ground is not frozen.
- Sow onions and leeks in seed trays in a heated propagator.
- Plant fruit trees and bushes, hedging and bare-root rose bushes.
- Chit seed potatoes.

Last quarter to new moon: 15th–21st

A dormant period, with low sap and poor growth. Do not sow or plant. A good time though for pruning, while sap is slowed. Weeding now will check growth well. Harvest any crops for storage. Fertilise and mulch the soil. Garden maintenance.

- Prune apple, pear, medlar and quince trees. Prune autumn-fruited raspberries, red and white currants, and gooseberries.
- Prune wisteria, cutting back long growths to 2–3 buds and avoiding flower buds.
- Clean and oil tools. Clean pots.
- Check your soil for its pH level. If it is low this would be a good time to add lime or calcified seaweed.
- Weed beds ahead of spring. Mulch areas that have not been recently limed with organic matter.

New moon to first quarter: 22nd–28th

The waxing of the moon is associated with rising vitality and upward growth. Towards the end of this phase plant and sow anything that develops above ground. Prepare for growth.

- You could sow chillies and aubergines, broad beans, peas and sweet peas now, in a heated propagator.
- Buy seeds and prepare seed trays or plugs and compost.
- Place forcers over rhubarb plants to exclude light and draw up the stems.

Note: Where no specific time for the change between phases is mentioned, this is because it happens outside of sensible gardening hours. For exact changeover times for any late-night or pre-dawn gardening, refer to the moon phase chart on page 19.

THE RECIPES

Bun of the month

Swiss Three Kings cake

Yeasted dough buns enriched with butter, sugar, eggs and sometimes chocolate and spices have long been eaten to mark moments of celebration throughout the year, the best known of these perhaps being hot cross buns. Every month of this almanac includes a recipe for a traditional bun with which to mark the season.

Three Kings cake, or *Dreikönigskuchen*, is primarily made in Switzerland to celebrate Epiphany on 6th January. It is usually flavoured with lemon and sultanas or chocolate chips. This version, however, contains dark chocolate chips and Seville orange zest and is given a sticky Seville orange glaze, to mark the arrival of Seville oranges in the shops this month. Traditionally, one of the buns would contain a small *fève* – a tiny porcelain figure or a dried bean – and the person who finds it is crowned king for the day.

Makes 1 large and 7 small buns
Ingredients
For the dough:
175ml milk, plus extra for egg wash
30g butter
50g brown sugar
300g strong white bread flour, plus extra for dusting
1 teaspoon salt
1 sachet (7g) instant yeast
50g hazelnuts, roughly chopped
50g dark chocolate, roughly chopped
Zest of 1 Seville orange
1 egg

For the glaze:

Juice and zest of 1 Seville orange

2 tablespoons brown sugar

Method

Line a large baking tray with parchment. Gently heat the milk, butter and sugar in a saucepan over a low heat until melted together; leave to cool. Combine the flour with the salt and yeast in a bowl. When the milk has cooled, add it to the flour mixture and form into a dough. Knead for a few minutes in the bowl, then add the chopped hazelnuts, dark chocolate and the orange zest. Knead to combine, then turn out onto a floured surface and knead for a further 5 minutes. Pop the dough back in the bowl, cover and allow it to puff up for 1 hour in a warm, dry place.

Meanwhile, preheat the oven to 200°C, Gas Mark 6. Separate the risen dough into 7 small balls and 1 larger one, and roll each in your hands until smooth. Place the big one in the middle of the baking tray and the little ones around it in a circle, being sure to leave enough room for the buns to puff out and touch in the second prove. Cover and leave to prove for up to 40 minutes in a warm, dry place. Brush the top of the buns with egg wash (the egg beaten with a little milk). Bake on the middle shelf for approximately 35 minutes.

While they bake, make the glaze. Combine the orange juice and zest with the brown sugar in a saucepan, allowing it to simmer for a couple of minutes. When the buns are fresh from the oven, brush with the glaze and serve.

Quick plough pudding

Plough Monday is the first Monday following Epiphany, and this year it falls on 9th January. It was once the beginning of the agricultural year and the day that farm workers would return to work following the Christmas season. In the 15th century a plough would be dragged through the streets and money collected for parish funds. This was then used to pay for 'plough lights', candles kept burning in church throughout this period to bless those working in the fields.

Plough pudding was a boiled suet pudding with a sausage filling and was traditionally served on Plough Monday. This recipe uses many of the same ingredients but is far quicker to throw together on a cold, dark Monday night in January, and the sausage can be replaced with a veggie version if you wish. The sage-and-onion stuffing mix gives this some 'retro roast dinner' flavours and helps to bind the whole thing together, but the dish also works well without it. It would go well with a sweet and spicy chutney, buttery mashed carrots and steamed greens.

Serves 4

Ingredients

6 large eggs

300ml milk

1 teaspoon salt

1 teaspoon pepper

300g white or brown bread, cubed

85g sage-and-onion stuffing mix

6 sausages, sliced into bite-sized rounds

| Cooking oil (optional) |
| 1 large onion, finely chopped |
| 1 Bramley apple, peeled and finely chopped |
| A few sprigs of thyme |
| 5 large sage leaves, finely sliced |
| 150g Gruyère cheese, finely grated |

Method

Preheat the oven to 180°C, Gas Mark 4 and generously grease a 23cm-square tin. In a large bowl, beat the eggs, milk and seasoning. Add the bread cubes and stuffing and mix together until evenly coated. Fry the sausage rounds in a large frying pan, adding a little oil if using veggie sausages. Add the onion, apple, thyme sprigs and sage, and fry for a few minutes until everything has a bit of colour and the ingredients release all their wonderful aromas.

Tip the fried ingredients into the bread mixture, and add most of the Gruyère, reserving a little for sprinkling on top. Stir with a large spoon until evenly distributed. The mixture should be quite wet, so add a splash more milk if necessary. Turn into the tin and press down a little. Sprinkle the remaining Gruyère over the top. Bake in the centre of the oven for about 40 minutes until it has browned and is puttering away nicely. Remove from the oven and allow to cool for 5 minutes before serving.

WHAT IS THE ZODIAC?

The zodiac is a band of sky that lies 8 degrees either side of the ecliptic, the path the sun takes through our sky during the year. The band is divided into 30-degree sections, each containing a constellation, and as the sun moves through the 12 constellations it moves in and out of the various zodiacal signs. The term 'zodiac' comes from the ancient Greek *zodiacus*, meaning 'cycle of little animals'. Astrologers believe that the time of our birth in relation to these signs affects our personalities, and that the movement of the planets of the solar system through the zodiac, and their aspects relative to each other, affects events, moods and movements on earth.

The Western zodiac is one of the oldest still in use, originating in Mesopotamia around 1900 years BCE. It spread through ancient Greece, Rome and the Arab world and finally made its way to western Europe. The 12 signs are well known to us – Aries, Taurus, Gemini, Cancer, Leo, Virgo, Libra, Scorpio, Sagittarius, Capricorn, Aquarius, Pisces – and each has a corresponding animal or symbol.

Jyotishya is traditional Hindu astrology, also known as Vedic astrology, and it, too, is based upon the idea of the sun moving through the constellations. Its *rasi* correspond pretty closely with Western zodiac signs – Mesa the ram coincides with Aries, Vrsabha the bull with Taurus, Mithuna the twins with Gemini – and it is thought that Jyotishya, too, may have been influenced by ancient Greek astrology. Astrology is still widely used in Hindu culture: it influences the names of babies and is consulted to find suitable dates for important days such as marriage and moving house.

Although the Chinese zodiac, or Sheng Xiao, is divided into 12 and has an animal assigned to each segment, it is not based on the movement of the sun through the constellations. Instead, each whole lunar year is assigned an animal and its characteristics in a 12-year cycle. People born in each year are believed to take on the sign's attributes. Years also alternate between Yin and Yang, Yin being dark/night/female/receptive and Yang being light/day/male/active, and each is assigned a 'fixed element' – water, earth, wood, fire or metal. Whereas this month's new moon falls on the 21st in the UK and Ireland, it falls on the 22nd in China, and this will mark the beginning of the new year, a Yin and water year, with Rabbit as its animal. Yin water is represented by mists, clouds and drizzle, and its colour is black, therefore this is the year of the Black Water Rabbit.

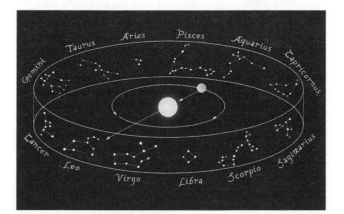

THE ZODIAC

Capricorn: 22nd December–19th January

The sun begins the month in the same area of sky that holds the constellation of Capricorn, the Goat, the 270th–300th degree of the zodiac. On the 20th of this month the sun will move into Aquarius (see page 50).

> **Symbol:** The Goat
> **Planet:** Saturn
> **Element:** Earth
> **Colour:** Brown
> **Characteristics:** Hard-working, diligent, ambitious, pragmatic, responsible

The Greek gods rewarded favoured mortals by placing them in the sky as constellations, and this is one of the stories behind the constellation of Capricornus (star sign Capricorn). The Titan Cronus was told that one day his own child would overthrow him, and so as each was born he ate them. When the sixth child, baby Zeus, was born, his mother, Rhea, handed Cronus a swaddled stone, which he quickly swallowed. Rhea hid Zeus in a cave on Mount Ida, in Crete, where he was suckled and raised by a goat, Amalthea. Sure enough, Zeus grew up to become king of the gods on Mount Olympus. When Amalthea died, her skin became the aegis – a magical protective garment burnished with gold that Zeus wore in battle – and Amalthea became a constellation, which some say was Capricornus. The best time to spot Capricornus is when it is in the opposite part of the sky from the sun six months from now, in July.

Also in astrology this month: Mercury is in retrograde this month, from 13th December 2022 to 18th January 2023, and astrologers believe this creates a period when communications break down, technology malfunctions, tempers fray and plans go awry (see page 82).

A FOLK SONG FOR CAPRICORN'S GOAT

'Bryan O'Lynn'
Traditional, arr. Richard Barnard

The folk songs in this year's almanac have links to our zodiac stories, and many of them provide an earthy, silly or downright bawdy change of tone from the grandiose posturing of the Greek myths. The link with this song is the goatskin – while Amalthea's skin after she died was burnished with gold and worn by Zeus in battle, Bryan O'Lynn's was made into a silly coat with the horns still attached. There is not much sense to this song, so don't look for it. The daftest and most nonsensical of British and Irish folk songs (and this was widely sung in England and Ireland) are often the oldest. 'Bryan O'Lynn' is thought to be around five hundred years old.

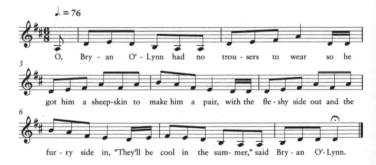

O, Bryan O'Lynn had no trousers to wear,
So he got him a sheepskin to make him a pair,
With the fleshy side out and the furry side in
'They'll be cool in the summer,' said Bryan O'Lynn.

O, Bryan O'Lynn had no shirt on his back,
So he went to the neighbours to borrow a sack.
He gathered the mouth of it under his chin
'They'll take them for ruffles,' said Bryan O'Lynn.

O, Bryan O'Lynn had no coat to put on,
So he got an old goatskin to make himself one.
With horns sticking out of his pockets, 'Well then,
'They'll answer for pistols,' said Bryan O'Lynn.

O, Bryan O'Lynn and his wife and wife's mother
They all three crossed over the river together.
The bridge it broke down and they all tumbled in
'We'll walk home on the bottom,' said Bryan O'Lynn.

NATURE

The pond in January

Perhaps the single most useful thing one person can do to help wildlife is to build a pond in their garden or allotment. There was once a 'dew pond' (a shallow pond, usually man-made, replenished with water from the dew) in the corner of every field for livestock to drink from, and they would be used not only by cows and sheep but by amphibians, insects, small mammals and birds. Most have been drained or filled in and, of those that remain, around 80 per cent are thought to be polluted or degraded, mainly by the nitrogen and phosphorus from agricultural fertiliser.

Fortunately, garden ponds are insulated from many of the problems that countryside ponds face, and it is remarkably straightforward to create a pond that quickly becomes an intricate ecosystem supporting dozens of species. The phrase 'build it and they will come' could have been written for ponds: just make one and sit back and you will see. This year's almanac follows the year in a garden pond, and all the wonderful things that are happening above and below the surface.

There are few signs of life in the January pond. Many garden birds visit it to drink and wash, and mammals will stop by to drink: during mild weather hedgehogs may even emerge from their sleep to take a drink. But other than that, all appears calm. Beneath the sometimes frozen surface there is life, but it is at its lowest ebb. The bottom of the pond is full of decaying sticks and leaves, and nestled within it are the larvae of beetles and insects, and even adult water beetles, which will occasionally return to the surface briefly to take in air. Nymphs of caddisflies, dragonflies and mayflies are down there and create a kind of antifreeze that prevents their bodies from freezing and their cells from rupturing. Dragonfly eggs nestled in the mud are in diapause, a type of hibernation that prevents them from hatching until the weather warms. Life is suspended, but not for long.

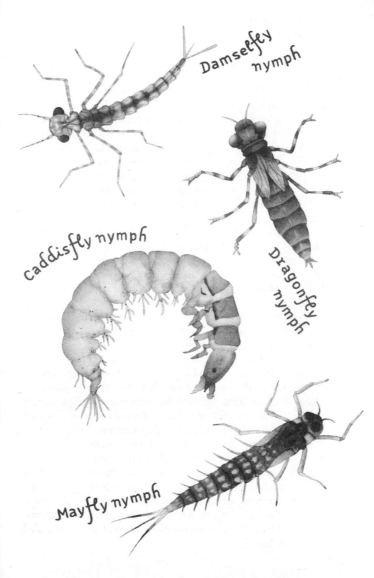

Damselfey nymph

Dragonfey nymph

Caddisfly nymph

Mayfly nymph

February

- **1** St Brigid's Day (Christian)
- **1** 1st–2nd: Imbolc (Gaelic/Pagan spring celebration)
- **1** Start of LGBT+ History Month
- **2** Candlemas (Christian)
- **6** St Brigid's Day bank holiday, Ireland
- **14** St Valentine's Day
- **15** Parinirvana Day/Nirvana Day (Buddhist)
- **20** Collop Monday/Peasen Monday/ Nickanan Night (Christian/traditional)
- **21** Shrove Tuesday/pancake day (Christian/ traditional)
- **22** Ash Wednesday – start of Lent (Christian)

FEBRUARY AT A GLANCE

In Scottish folklore, Beira, sometimes known as the Cailleach, is the goddess of winter, and she can cause a frost by banging her staff hard on the ground. She holds a young woman, Bride or Brigid, captive in the icy northern kingdom, and one day sends her to a freezing stream with a brown cloak, instructing her to wash it white, which of course she cannot. Bride despairs and is rescued by Father Winter, who turns the cloak white and presents her with a bunch of snowdrops. When she returns with them, Beira is enraged. The snowdrops are a sign that her power is waning, and she sets off across the land, banging her staff and bringing frosts and snowfall wherever she goes.

It is snowdrops that encapsulate the energy of February, pushing up through the cold ground, sometimes beneath a blanket of snow. They are a sign that things are moving towards spring, but the story provides a warning in the shape of that icy staff: there is plenty more winter still to come.

We get a great boost this month from the lengthening of the evenings, and a few mild days can see us itching to start the year, to sow seeds and throw ourselves into action. We are like shoots pushing through the dark soil, faces eagerly towards the sun. But it is too soon. The Gaelic festival of Imbolc falls on the 1st–2nd, and the word may come from *Oímelc*, the Old Irish word for the beginning of spring (itself deriving from *oí-melg*, meaning 'ewe's milk' or 'in the belly', a reference to pregnancy and the forthcoming lambing). This moment is that initial stirring into life, first pushing tentatively above the soil. Make plans and gather seeds for future sowing, but tend your flame carefully through these icy days and long dark nights.

THE SKY AT NIGHT

There will be a good show of planets this month, but Saturn is now lost as an evening star, and Mars is dimming.

22nd: Venus and Jupiter close to the moon. They first appear in the dusk at around 18.00 at an altitude of 24 degrees in the southwest. They set in the west at around 20.00.

27th: Close approach of the moon and a dim Mars. First visible in the dusk at around 18.00 at an altitude of 64 degrees in the south. They then get closer together until they set in the northwest at around 02.40 the next day.

THE SOLAR SYSTEM

Uranus

For our second planet of the year we head deep into the cold and murky further reaches of the solar system to the seventh planet from the sun and the first of the Ice Giants, Uranus. Its link to February is via this month's star sign, Aquarius, which is said to be ruled by Uranus, the god of the sky, who was father to Cronos and grandfather to Zeus. (The Roman equivalents of the Greek Cronus and Zeus are Saturn and Jupiter.)

The blue-green surface that we see when we look at photos of the planet Uranus is not solid. It is a mass of hydrogen, helium and ice, and is almost featureless, perfectly smooth and without the cloud systems that feature on the surfaces of other gas giants. Uranus moves incredibly slowly, taking 84.3 years to make one circuit around the sun. It also lies nearly on its side, its pole almost in line with the plane of the solar system, therefore each pole experiences around 42 years of light and then 42 years of darkness.

In theory Uranus can be seen with the naked eye, but it is very faint and you would need a perfectly dark sky. It will be at opposition – and so at its closest and brightest – in November, so that would be the time to try to spot it.

Sunrise and set
Haltwhistle, Northumberland

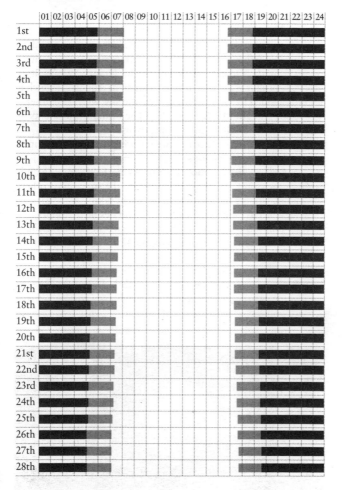

THE SEA

Average sea temperature in Celcius

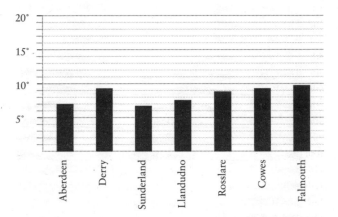

Spring and neap tides

Spring tides are the most extreme tides of the month, with the highest rises and the lowest falls, and they follow a couple of days after the full moon and new moon. These are the times to choose a low tide and go rock-pooling, mudlarking or coastal fossil-hunting. Neap tides are the least extreme, with the smallest movement, and they fall in between the spring tides.

Spring tides: 7th–8th and 21st–22nd

Neap tides: 1st–2nd and 13th–14th

Spring tides are shaded in black in the chart opposite.

February tide timetable for Dover

For guidance on how to convert this for your local area, see page 8.

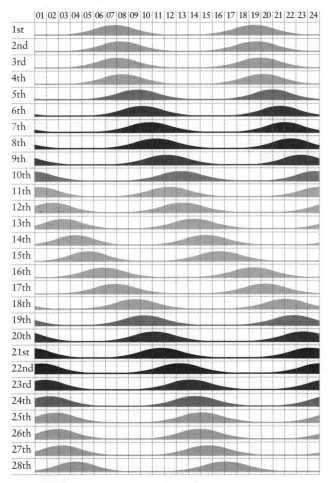

THE MOON

Moon phases

Full moon – 5th February, 18.29

Last quarter – 13th February, 16.01

New moon – 20th February, 07.06

First quarter – 27th February, 08.06

Moonrise and set

Like the sun, the moon rises roughly in the east and sets roughly in the west. It also rises around 50 minutes later each day. Use the following guide to work out approximate moonrise times.

Full moon: Rises near sunset, opposite the sun, so in the east as the sun sets in the west.
Last quarter: Rises around midnight, and is at its highest point as the sun rises.
New moon: Rises at sunrise, in the same part of the sky as the sun (and so cannot be seen).
First quarter: Rises near noon, and is at its highest point as the sun sets.

Full moon

February's full moon is known as the Snow Moon, Ice Moon or Storm Moon.

New moon

This month's new moon, on the 20th, is in Aquarius. Astrologers believe that the new moon is a quiet, contemplative time before a phase of growth. Each new moon has its own energy, depending on the zodiacal sign that it is in, and the Aquarius new moon is said to rule our intuition, innovation and psychic state.

Moon phases for February

1st	2nd	3rd	4th	5th FULL

6th	7th	8th	9th	10th

11th	12th	13th	14th	15th

16th	17th	18th	19th	20th NEW

21st	22nd	23rd	24th	25th

26th	27th	28th		

GARDENS

To enjoy this month

Ornamental: Snowdrops, camellias, *Daphne odorata*, crocuses, primulas, hellebores, stinking hellebore, aconites, hazel catkins, goat willow catkins, *Iris reticulata*, sweet violets, primroses, hyacinths, blackthorn blossom
Edible: Forced rhubarb, purple sprouting broccoli, Oriental leaves, carrots, leeks, parsnips, swedes, Jerusalem artichokes, chervil, coriander, parsley, rosemary, sage, bay

Gardening by the moon
The following is a guide to planting with the phases of the moon, according to traditional practices. It also works as a guide to the month's gardening for moon-gardening cynics, who can do these jobs whenever they wish during the month ahead.

First quarter to full moon: 29th January–5th and 27th–7th March (till 12.40)
Sow crops that develop above ground, don't sow root crops. Plant out seedlings and young plants. Take cuttings and make grafts. Avoid any other pruning. Fertilise.
- Sow early lettuces, winter salad leaves, spinach, radishes, hardy peas and early varieties of Brussels sprouts, kohlrabi and sprouting broccoli in pots under cover.
- Sow broad beans straight into the ground if it is not frozen and cover them with cloches.
- Sow chillies and aubergines indoors in a heated propagator.

Full moon to last quarter: 6th–13th
A 'drawing down' energy. This phase is a good time for sowing and planting any crops that develop below ground: root crops, bulbs and perennials. Light is high but decreasing.
- Plant Jerusalem artichokes, garlic, rhubarb crowns and shallot and onions sets if the ground is not frozen.
- Sow onions and leeks in seed trays in a heated propagator.

- Plant fruit trees and bushes, grapevines, hedging and rose bushes.
- Chit seed potatoes.

Last quarter to new moon: 14th–19th

A dormant period, with low sap and poor growth. Do not sow or plant. A good time though for pruning, while sap is slowed. Weeding now will check growth well. Harvest any crops for storage. Fertilise and mulch the soil. Garden maintenance.

- Prune apple, pear, medlar and quince trees
- Prune autumn-fruited raspberries, red and white currants and gooseberries.
- Prune wisteria, cutting back long growths to 2–3 buds and avoiding flower buds.
- Clean and oil tools. Clean pots.
- Weed beds ahead of spring. Mulch them with organic matter. Prepare for planting out.
- Pin clear or black polythene over beds to warm the soil for early plantings.

New moon to first quarter: 20th–26th

The waxing of the moon is associated with rising vitality and upward growth. Towards the end of this phase plant and sow anything that develops above ground. Prepare for growth.

- Sow chillies and aubergines, broad beans and peas, or wait for the first quarter phase, which is more suited to sowing.
- Buy seeds and prepare seed trays or plugs and compost. Get your propagator set up.
- Place forcers over rhubarb plants to exclude light and draw up the stems. Give spring cabbages and other brassicas a high-nitrogen feed.

Note: Where no specific time for the change between phases is mentioned, this is because it happens outside of sensible gardening hours. For exact changeover times, refer to the moon phase chart on page 43.

THE RECIPES

Bun of the month

Rhubarb and cardamom cream *semlor*

In Sweden, *semlor* (singular: *semla*) are eaten on Fat Tuesday/
Shrove Tuesday, the day before the end of Shrovetide and
the beginning of Lent. The idea is to have a festival of dairy
and sugar before the privations of the weeks ahead, but at
some point this got upended and now Swedes eat *semlor* on
every Tuesday between Fat Tuesday and Easter. They are also
eaten in Finland, where they are known as *fastlagsbullar* (Fat
Tuesday rolls; singular: *fastlagsbulle*). In their traditional
incarnation they are spiced with cardamom and are split and
filled with sweetened almond paste and whipped cream, but
in Finland jam sometimes replaces the almond paste. The
version given here substitutes beautifully pink poached forced
rhubarb, which is briefly appearing in the shops now. You
could also swap the cardamom cream for custard to make
rhubarb and custard *semlor*.

Makes 12
Ingredients
100ml milk
150ml water
100g butter
500g strong white bread flour, plus extra for dusting
80g caster sugar
1 sachet (7g) instant yeast
2 teaspoons ground cardamom
1 egg

For the filling:

250ml double cream

1 teaspoon vanilla extract

1 teaspoon ground cardamom

2 tablespoons icing sugar, plus extra to serve

3 sticks rhubarb, diced

1 tablespoon granulated sugar

Method

Line a baking tray with parchment. In a saucepan heat the milk, measured water and butter until melted; leave to cool. Combine the strong white bread flour, caster sugar, yeast and ground cardamom in a separate bowl. Make a well in the centre and crack the egg into it, then mix in the cooled milk mixture, to make a sticky dough. Add a little more flour – just enough to stop it from sticking to the surface but not too much, as you need to keep it light. Turn out the dough on a floured board, and knead until it is smooth and elastic, then pop it back in the bowl, cover and leave in a warm place for about an hour. Once the dough has doubled, turn it out and divide into 12 portions. Roll each into a smooth ball between your hands and place on the baking tray. Cover and allow to prove again for around 45 minutes.

Meanwhile, preheat the oven to 200°C, Gas Mark 6. When the buns have risen, bake them for 12–15 minutes. Whip the double cream with the vanilla extract, ground cardamom and icing sugar. Soften the rhubarb in a saucepan with the granulated sugar for 10 minutes on a low heat. When the buns have cooled, slice off the top of each one with a sharp knife, scoop out a little of the bun, then add some whipped cream mixture and a spoonful of the rhubarb. Balance the 'hat' on top of each, and sprinkle with a little more icing sugar to serve.

Pea soup with cheesy mushroom dumplings

Nickanan Night is a Cornish name for the Monday before Lent (also known as Collop Monday, Peasen Monday or any number of other names). In Cornwall this has traditionally been a night for mischief, when gangs of boys would roam around knocking on doors and running away, a practice known as 'nicky nicky nine doors'. Pea soup was eaten on this night – the peas, of course, being dried split peas rather than fresh ones – and often with a dish of bacon and eggs in these last days before the Lenten period, which would prohibit such things.

Serves 4

Ingredients

For the soup:

200g yellow split peas

1.7 litres water

1 large onion, roughly chopped

4 bay leaves

1 star anise

1 teaspoon salt

1 tablespoon balsamic vinegar

For the dumplings:

20g dried mushrooms

125g self-raising flour

Pinch of salt and pepper

75g butter

40g Gruyère or Cheddar cheese, finely grated

20g sunflower seeds

Method

Soak the split peas in cold water for at least 8 hours, then drain and put them in a large saucepan with the measured water, onion, bay leaves and star anise. Bring to the boil, skimming off any foam, then simmer for 2 hours, stirring occasionally.

Meanwhile, make the dumplings. Soak the dried mushrooms in boiling water so they are covered and leave to rehydrate for 20 minutes. Once plumped up, drain the mushrooms, reserving the water, and chop the mushrooms finely. Put the flour, salt and pepper in a bowl, then grate in the butter. Use your fingertips and thumbs to lift and rub the butter into the flour for a minute until it resembles crumbs. Tip in the cheese, mushrooms and sunflower seeds, and stir to combine. Using a little of the mushroom water, bring the mixture together into a soft dough and form into 8 balls. (If you don't need all of them, they will freeze well in this uncooked state, for addition to future soups.)

When the soup is looking thick and a little cloudy, add the salt and the vinegar, taste and add more if needed. Turn down to the lowest heat and add the dumplings, leaving a bit of space between each ball. Put the lid on and cook for 30 minutes. Remove the bay leaves and star anise, and serve.

THE ZODIAC

Aquarius: 20th January–18th February

The sun begins the month in the same area of sky that holds the constellation of Aquarius, the Water Bearer, the 300th–330th degree of the zodiac. On the 19th of this month it will move into Pisces, see page 73.

> **Symbol:** The Water Bearer
> **Planet:** Uranus
> **Element:** Air
> **Colour:** Turquoise
> **Characteristics:** Independent, enigmatic, socially conscious, idealistic, original, easy-going.

Young Ganymede was the most beautiful of all mortals. One day he was tending his sheep on Mount Ida near Troy in Phrygia (now part of modern Turkey), the soft wind wafting through his dark, glossy locks and the sun falling across his perfectly sculpted nose, when Zeus saw him and fell instantly in love with him. Zeus turned himself into an eagle and stole Ganymede away to Mount Olympus. There Zeus made Ganymede the cupbearer (or water bearer/wine pourer) to the gods, and granted him eternal youth and immortality. Later he placed him in the sky, as Aquarius, the Water Bearer, where he stands alongside the constellation of Aquila, the Eagle.

The best time to spot Aquarius is when it is in the opposite part of the sky from the sun six months from now, in August.

A FOLK SONG FOR AQUARIUS' WATER BEARER

'The Sussex Toast'
Traditional, arr. Richard Barnard

A song to tie in with the story of Ganymede, cupbearer to the gods, and who the constellation Aquarius represents. Similar versions of this song were once found throughout Sussex but also in Herefordshire and Devon. It was likely a comic drinking song performed by a group drinking together in a pub; the solo singer points out and mocks particular individuals in the group with each verse, and the group joins in for the last line. Best sung rowdily.

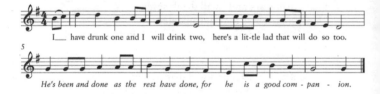

I have drunk one and I will drink two,
Here's a little lad that'll do so too.
He's been and done as the rest have done,
For he is a good companion.

I have drunk two and I will drink three,
Here's a little lass that'll drink like me.
She's been and done as the rest have done,
For she is a good companion.

I have drunk three and I will drink four,
Here's a little lad that has double his score.
He's been and done as the rest have done,
For he is a good companion.

I have drunk four and I will drink five,
Here's a little lad that drinks dead or alive.
He's been and done as the rest have done,
For he is a good companion.

I have drunk five and I will drink six,
Here's a little lad in a fair good fix.
He's been and done as the rest have done,
For he is a good companion.

I have drunk six and I will drink seven,
Here's a little lass that'll drink eleven.
She's been and done as the rest have done,
For she is a good companion.

I have drunk seven and I will drink eight,
Here's a little lad that'll drink his straight.
He's been and done as the rest have done,
For he is a good companion.

I have drunk eight and I will drink nine,
Here's a little lad that'll drink till he's blind.
He's been and done as the rest have done,
For he is a good companion.

I have drunk nine and I will drink ten,
Here's a little lad that'll say 'Amen'.
He's been and done as the rest have done,
For he is a good companion.

NATURE

The pond in February

After a winter of peace and quiet in the pond, it will leap into riotous, libidinous life in February. Not at the beginning of the month though: peering into the pond you could be forgiven for thinking that nothing much is happening down there in the still, cold depths. Plant growth is beginning, though. The shoots of bulrush, or reed mace, begin to vigorously push up out of the water early in the year. They were traditionally harvested on St Brigid's Eve, 31st January, and the leaves made into crosses with woven square centres for St Brigid's Day on 1st February, when they were hung over doors to protect the home from harm.

But the main attractions this month are the frogs and toads. A slight increase in temperature will signal to the frogs hibernating in and around the pond that it is time to wake up and begin feeding on the grubs and insects in the pond to fatten up. Toads are more adapted to life on land than frogs, but they must return to their 'home pond' to breed, and this, across much of the UK and Ireland, is the month they move, many making long and dangerous migrations, risking their lives crossing roads and whatever obstacles are in their way.

And all to partake in a great frenzy of breeding late in the month. Both toads and frogs take to the pond, with the male climbing onto a female's back and holding her tight in a sexual embrace known as amplexus, his front legs grasped around her to hold on, while he uses his back legs to shove away the many competing males. The male is actually just holding the female in an embrace so that he will be the closest to her when she lays her eggs into the water, which is when he will deposit his sperm and fertilise the eggs. This breeding season lasts 12–24 days but there is a climax of 3–7 days when the males' night chorus – a sort of purring noise, with croaks and whistles – reaches its height, often under a full moon.

Common frog

Pool frog

Frogspawn

Common toad

Natterjack toad

Toadspawn

March

 Start of meteorological spring

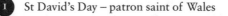 St David's Day – patron saint of Wales

 St Piran's Day – patron saint of Cornwall

 Holi (Hindu spring festival) begins at sundown

 Purim (Jewish) begins at sundown

 International Women's Day

 St John of God's Day – patron saint of booksellers

 St Patrick's Day – bank holiday, Northern Ireland, Ireland

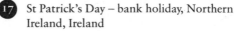 Mothering Sunday/fourth Sunday in Lent (traditional/Christian)

Vernal equinox, at 21.24 – start of astronomical spring

Ostara – Pagan celebration of spring

Ramadan (Islamic month of fasting, prayer and reflection) begins at sundown/sighting of crescent moon

 British Summer Time (BST) and Irish Standard Time (IST) begin – both are Greenwich Mean Time/Coordinated Universal Time + one hour. Clocks go forward one hour at 01.00

MARCH AT A GLANCE

A moment of perfect balance will occur this month. On the 20th of March at 21.24 the sun will be exactly above the equator. Pole to pole, the earth will be half in shadow, half in sunlight. The whole world will have 12 hours of day and 12 hours of night: Norway, Jamaica, Brazil, New Zealand and the UK and Ireland.

We are all looking for balance in our lives, but March, with the vernal/spring equinox, is a good moment to think about the nature of it. Balance is not a stable state. Mere moments later Norway will start to stretch towards summer, and New Zealand will shrink towards winter. That moment of balance is brief in the extreme, a tiny sliver of time. Like standing on one leg, if we want more balance in our lives we have to work at it, infinitesimally adjusting to change, concentrating our minds.

By the end of the month we will have a tiny bit more day than we have night, and that will be increasing all the time. We will be into the light half of the year, and the desire to fling ourselves at the world is great. The energy for new projects is high and life feels sunny and buzzy, all daffodils and bumblebees and miraculously lengthening evenings. But think forward now to the autumn equinox. Imagine reaching the end of summer and not feeling frazzled and burnt out and so desperately in need of comfort food and peace. Imagine taking some of the most helpful lessons of the dark half of the year – how to cocoon, how to soothe yourself, how to rest – and bringing them with you now through the bright half of the year.

THE SKY AT NIGHT

There are some nice close approaches coming up. Jupiter will
be lost in the glare of the sun by the end of this month, and
Mars is very dim, but Venus is now getting higher and brighter
in the evening sky.

2nd: Close approach of Venus and Jupiter. First visible in
the dusk at around 18.00 at an altitude of 22 degrees in the
west. Jupiter will be 0.7 degrees below Venus: very close.
They will go on to set in the west at about 20.10.
24th: Close approach of the moon and Venus. First visible
in the dusk at around 18.40 at an altitude of 27 degrees in
the west. They will then go on to set in the northwest at
about 21.20.
27th: Close approach of the moon and a dimming Mars.
They first appear in the dusk at about 19.40 in the south
at an altitude of 62 degrees. They go on to set in the
northwest at around 02.50 the next day.

THE VERNAL/SPRING EQUINOX

The earth takes 24 hours to rotate on its axis, with half of the earth in shadow (night) and half in light (day). It does this on an angle of 23.4 degrees known as the 'axial tilt', and it is this angle that is responsible for our seasons. Because as well as rotating on its own axis, the earth also orbits the sun, taking a year to do so. During this time the axial tilt stays constant, and so, as the earth moves around the sun, different parts receive greater or smaller amounts of sun.

When the earth is at the point where the north pole is leaning towards the sun, we have reached the northern hemisphere's summer solstice, and the southern hemisphere's winter solstice. When the earth moves around to the other side of the sun, so that the south pole is tilted towards the sun, then the southern hemisphere has its summer and the northern hemisphere has its winter. The equinoxes, in spring and autumn, occur at exactly the halfway points between these moments. The axis of the earth at these times is side-on to the sun, which means that – momentarily – no hemisphere is favoured. Day and night are roughly even, all over the world, from tropical Colombia at the equator to icy Lapland in the Arctic Circle. The exact moment of the equinox is when the sun is directly overhead at the equator.

Common wisdom has it that, at the equinox, day and night are exactly the same length, but this is not quite true. It would be correct if sunrise and set were measured from the moment that the centre of the sun rises over or sinks below the horizon, but we measure them from the moment the top edge appears or disappears. Thus, the equinox day is a little longer than its night, and its length varies slightly through the world. In spring, 'equilux', which is when day and night are exactly the same length, occurs a few days before the equinox.

Naming of the equinox
The word 'equinox' is derived from two Latin words, *aequus* ('equal') and *nox* ('night'). The spring equinox is

M

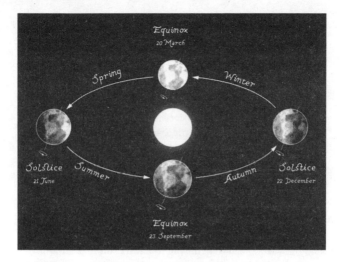

sometimes called the vernal equinox, the word vernal being
a 16th-century English word derived from the Latin *vernalis*,
which itself is derived from the Latin word for spring, *ver*.
The Welsh name for the spring solstice is Alban Eilir, *alban*
meaning 'quarter year' and *eilir* meaning 'spring'. In Scottish
Gaelic it is called *co-fhad-thrath an earraiche* and in Irish
Gaelic *conocht an earraiche*. Spring equinox is called Ostara
by modern Wiccans, Druids and Pagans, after the Germanic
goddess Eostre, who is associated with the east and dawn.

Vernal/spring equinox facts

Date: 20th March
Time: 21.24
Altitude: At solar noon on 20th March, the sun will reach 38
degrees in the London sky, 34 degrees in the Glasgow sky and
37 degrees in the Dublin sky.
Sunrise times: London 06.04, Glasgow 06.20, Dublin 06.28
Sunset times: London 18.13, Glasgow 18.30, Dublin 18.37
Sun's distance from earth: 148,965,000 kilometres

Mark the vernal/spring equinox

- Go outside at dawn and greet the sun. Think about the dawn stretching from pole to pole, and about the earth's position, side-on to the sun as it makes its annual orbit around it. Feel the warmth of the sun on your face, knowing that it is getting stronger.
- Gather together materials for a nature table or an altar to mark the moment. Include a small vase of spring flowers, unfurling buds and blossom, some moss formed into a nest, a few found feathers and some candles. Light the candles at 21.24.
- Fill a spray bottle with water, 1 teaspoon of vodka and 25–30 drops of spring-like essential oils, then shake it up and spritz it around. Try: lemon, jasmine, grapefruit and chamomile.

Sunrise and set
Haltwhistle, Northumberland

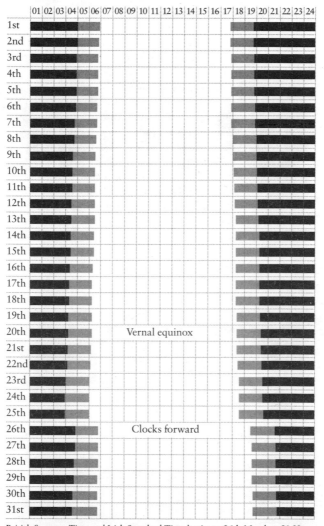

British Summer Time and Irish Standard Time begin on 26th March at 01.00
and this has been accounted for above.

THE SEA

Average sea temperature in Celcius

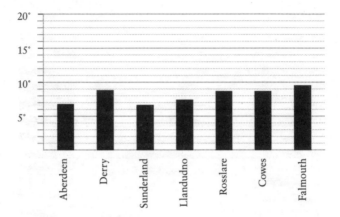

Spring and neap tides

Spring tides are the most extreme tides of the month, with
the highest rises and the lowest falls, and they follow a couple
of days after the full moon and new moon. These are the
times to choose a low tide and go rock-pooling, mudlarking
or coastal fossil-hunting. Neap tides are the least extreme,
with the smallest movement, and they fall in between the
spring tides.

Spring tides: 8th–9th and 22nd–23rd

Neap tides: 1st–2nd, 16th–17th and 30th–31st

Spring tides are shaded in black in the chart opposite.

March tide timetable for Dover

For guidance on how to convert this for your local area, see page 8.

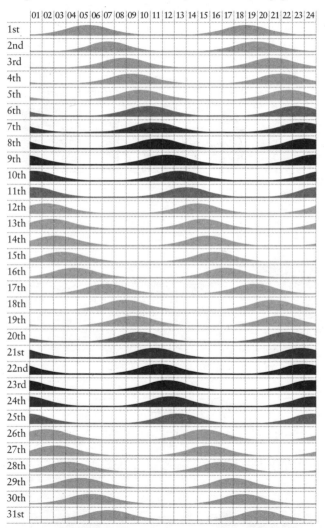

THE MOON

Moon phases

Full moon – 7th March, 12.40

Last quarter – 15th March, 02.08

New moon – 21st March, 17.23

First quarter – 29th March, 03.32

Moonrise and set

Like the sun, the moon rises roughly in the east and sets roughly in the west. It also rises around 50 minutes later each day. Use the following guide to work out approximate moonrise times.

Full moon: Rises near sunset, opposite the sun, so in the east as the sun sets in the west.
Last quarter: Rises around midnight, and is at its highest point as the sun rises.
New moon: Rises at sunrise, in the same part of the sky as the sun (and so cannot be seen).
First quarter: Rises near noon, and is at its highest point as the sun sets.

Full moon

March's full moon is known as the Plough Moon, Lenten Moon or Chaste Moon.

New moon

This month's new moon, on the 21st, is in Pisces. Astrologers believe that the new moon is a quiet, contemplative time before a phase of growth. Each new moon has its own energy, depending on the zodiacal sign that it is in, and the Pisces new moon is said to rule spirituality and interconnection.

Moon phases for March

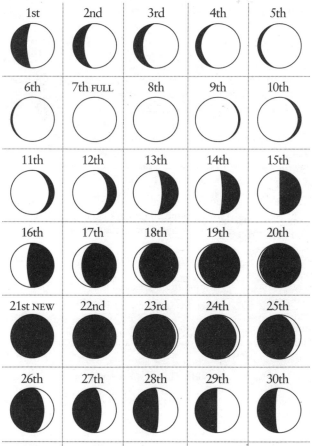

GARDENS

To enjoy this month

Ornamental: Daffodils, camellias, fritillaries, primulas, stinking hellebore, magnolias, wallflowers, pussy willow, amelanchier, scilla, euphorbia, cherry blossom, blackthorn blossom
Edible: Wild garlic, nettle tips, forced rhubarb, purple sprouting broccoli, Oriental leaves, leeks, parsnips, swedes, Jerusalem artichokes, chervil, coriander, parsley, rosemary, sage, bay

Gardening by the moon
The following is a guide to planting with the phases of the moon, according to traditional practices. It also works as a guide to the month's gardening for moon-gardening cynics, who can do these jobs whenever they wish during the month ahead.

First quarter to full moon: 27th February–7th (until 12.40) and 29th–5th April
Sow crops that develop above ground, don't sow root crops. Plant out seedlings and young plants. Take cuttings and make grafts. Avoid any other pruning. Fertilise.
- Sow aubergines, chillies and peppers, cucumbers and tomatoes indoors in a heated propagator.
- Sow Brussels sprouts, summer and autumn cabbages, celery, Florence fennel, lettuces and sprouting broccoli in pots or modules under cover.
- Sow hardy annual flower seeds.
- In mild areas and on light soils or ground that has been pre-warmed by covering, direct sow lettuces, peas, spinach and Swiss chard, and cover with cloches.
- Plant out broad beans and peas, under cloches.
- Gradually increase feeding and watering of plants.

Full moon to last quarter: 7th (after 12.40)–14th
A 'drawing down' energy. This phase is a good time for sowing and planting any crops that develop below ground: root crops, bulbs and perennials. Light is high but decreasing.

M

- Plant onions sets, rhubarb crowns, asparagus crowns and Jerusalem artichokes.
- Plant new cold-stored strawberry runners and sow seeds of alpine strawberries.
- Lift, divide and replant perennial herbs.
- Chit seed potatoes.
- Start planting out first early potatoes.
- In mild areas sow carrots, turnips, beetroot and radishes, and cover with cloches.
- Lift, split and replant crowded clumps of snowdrops.
- Plant lilies, dahlias, gladioli.

Last quarter to new moon: 15th–21st
A dormant period, with low sap and poor growth. Do not sow or plant. A good time though for pruning, while sap is slowed. Weeding now will check growth well. Harvest any crops for storage. Fertilise and mulch the soil. Garden maintenance.

- Prune raspberries, red and white currants, and gooseberries.
- Weed and mulch beds and prepare them for planting out.
- Protect cherry, apricot, peach and nectarine blossom from frosts. Plastic/glass coverings will protect peaches and nectarines from peach leaf curl.
- Feed perennials and overwintering plants – such as onions, kale, cabbages, hardy lettuces and leaves – with liquid feed.
- Feed and mulch fruit trees and bushes.
- Make a bean trench: dig out a trench, line it with newspaper and fill with compost and organic waste matter.

New moon to first quarter: 22nd–28th
The waxing of the moon is associated with rising vitality and upward growth. Towards the end of this phase plant and sow anything that develops above ground. Prepare for growth.

Note: Where no specific time for the change between phases is mentioned, this is because it happens outside of sensible gardening hours. For exact changeover times, refer to the moon phase chart on page 67.

THE RECIPES

Bun of the month

Cornish sweet–savoury saffron buns

A bun to mark St Piran's Day, the feast day of the patron saint of Cornwall, on 5th March. Saffron is believed by some to have found its way to Cornwall from Greece and the Middle East as early as 400 BCE, as a trading commodity for tin, which has been mined in Cornwall since 2150 BCE. Certainly saffron has been imported into Cornwall since the 14th century CE. Saffron is often combined with raisins, mixed peel and sugar to make sweet teatime buns, golden in colour, to be split and toasted. Here it is combined with pine nuts and dates in a nod to its Middle Eastern ancestry, to make a bun that can swing both ways, with a sweet but slightly savoury bent. Serve it fresh with butter or try it toasted with blue cheese and fig jam.

Makes 12
Ingredients
300ml milk
140g butter
1 tablespoon clear honey
Pinch of saffron strands
300g strong white bread flour, plus extra for dusting
300g spelt flour
1 teaspoon salt
1 sachet (7g) instant yeast
50g pine nuts, lightly toasted
50g dates, finely chopped

For the glaze:

1 tablespoon clear honey

5 tablespoons water

1 teaspoon ground coriander

Method

Line a baking tray with parchment. Gently heat the milk, butter and honey in a saucepan until melted together, then remove from the heat. Add a good pinch of saffron strands and allow to cool and infuse. Combine the strong white flour, spelt flour, salt and yeast in a big bowl. Make a well and pour in the cooled liquid, bringing the ingredients together into a dough. Turn out onto a floured surface and knead until smooth and elastic. If the dough is sticky, add a little extra flour. Add the pine nuts and dates, and knead for 5 minutes. Pop the dough back in the bowl, cover and leave in a warm place for about 1 hour, until double in size.

Knock back the dough and divide into 12 balls, rolling each piece in your hands or on a surface. Place on the baking tray, cover and allow to rise again for 45 minutes. Meanwhile, preheat the oven to 200°C, Gas Mark 6, and when the buns have risen, bake them for 20 minutes.

Make a glaze by boiling the honey, measured water and ground coriander in a saucepan for 1 minute. When the buns come out, brush the glaze over the tops until evenly covered.

Wild garlic salt

Woods are full of wild garlic now – find it by following your nose. You can, of course, gather it up and chop it into omelettes, risottos, pestos and breads straight away, but you can also extend its brief season by preserving a little in a flavoured salt, which will bring that fresh, herbal, gently garlicky flavour to future dishes, too.

Makes 3–4 jars
Ingredients
100g wild garlic
1kg sea salt flakes

Method

Wash and dry the wild garlic and then process into a purée, either with a mortar and pestle or in a food processor. Add 100g of the salt and process again. Put the rest of the salt into a large bowl, add the purée and stir it in with a fork to break it up and spread it through the salt.

When it has all turned vibrant green, spread it out on a baking tray and leave it to dry in the sun under a fine mesh for at least 5 hours, or in a very low oven for about 2 hours. When dry and cool, tip into sterilised jars. It will keep for a few months.

THE ZODIAC

Pisces: 19th February–20th March

The sun begins the month in the same area of sky that holds
the constellation of Pisces, the Fishes, the 330th–360th degree
of the zodiac. On the 21st of this month the sun will move
into Aries (see page 94).

> **Symbol:** The Fishes
> **Planet:** Neptune
> **Element:** Water
> **Colour:** Sea green
> **Characteristics:** Creative, intuitive, intelligent, sensitive,
> non-judgemental, spiritual

Typhon, the most terrifying of the monsters, had been sent
by Gaia to attack the gods. To escape, Aphrodite and her
son Eros escaped by leaping from the banks of the Euphrates
onto the backs of two fish, which bore them away. Aphrodite
put the fish into the sky, where they became the constellation
Pisces. The best time to spot Pisces is when it is in the
opposite part of the sky from the sun six months from now,
in September.

A FOLK SONG FOR PISCES' FISHES

'The Herring Song'
Traditional, arr. Richard Barnard

A fishing song for the fish of Pisces, this month's zodiac sign. It was sung all around the coast of England, wherever men fished for herring. There are many versions, including those from Cornwall, Devon, Northumbria and East Anglia, where it was known as 'The Yarmouth Herring'. This herring might even be as magical as the ones in the Pisces myth, with its great number of uses, its fins turning to needles and its tail to a sail for a ship.

M

O, what will I do with my herring's head?
I'll make a good oven as ever baked bread;
Ovens and bread and other things too,
But still there is more with my herring to do.

O, what will I do with my herring's eyes?
We'll cook them all up into puddings and pies;
Puddings and pies and other things too,
But still there is more with my herring to do.

O, what will I do with my herring's ribs?
Make forty new cradles and fifty new cribs;
Cradles and cribs and other things too,
But still there is more with my herring to do.

O, what will I do with my herring's fins?
I'll make them all into sharp needles and pins,
Needles and pins and other things too,
But still there is more with my herring to do.

And what do I do with my herring's tail?
I'll make the best ship that I ever saw sail;
Ships and sails and all sorts of things,
And now there's no more of my herring to sing.

NATURE

The pond in March

The frogs and toads started the party last month, but the rest of the pond – plants, bugs and all – starts to come alive this month. The remnants of the night before are everywhere. Much of the surface is now covered in shiny clumps of frog spawn and strings of toad spawn. There is always way too much for all of the frogs to hatch, but nature will work itself out and there is no need to move any of it. The vast numbers are insurance against frost and predation, and many of the eggs will be eaten, providing sustenance for other creatures in the pond. Newts feast on the eggs, leaving the empty casings behind, ahead of their own egg-laying, always well behind the frogs and the toads.

By the end of this month those that have survived the first month will have hatched, and the masses of wriggling tadpoles will be feasting on the pond's algae and vegetation, clinging on and wiggling their tails. Frog tadpoles are slim and grey, flecked with gold, while toad tadpoles are chunkier and pure black. Dragonfly and damselfly nymphs will hunt and feast on the tadpoles. They have a lot of growing to do, and will moult up to 17 times before they are fully grown. Juvenile backswimmers and water beetles eat tadpoles, too – it's brutal, but all part of the great circle of life that is the garden pond.

The plants around the pond and in it begin to creep back to life in March. Marginal plants such as brooklime and water forget-me-not start to green up, putting on fresh stems and leaves. Arrowhead emerges from the water, and duckweed starts to spread across the surface (pull it out by hand when you see it, as it will quickly cover it), but the big burst of foliage and flowers will come once the weather has warmed up more.

Lengthening days also trigger the start of breeding for a great many of the pond's tinier creatures now. The breeding season has begun for pond snails, backswimmers, boatmen, diving beetles, whirligig beetles and water mites. The pond year is up and running.

M

Water milfoil

Willow moss

Curly pondweed

April

1 April Fools' Day

2 Palm Sunday (Christian)

5 Passover/Pesach (Jewish) begins at sundown, with the Seder feast

7 Good Friday (Christian) – bank holiday, England, Scotland, Wales, Northern Ireland

9 Easter Sunday (Christian)

10 Easter Monday (Christian) – bank holiday, England, Wales, Northern Ireland, Ireland

14 St Tiburtius' Day – traditionally cuckoos start singing today (and stop on St John's Day, 24th June)

14 Orthodox Good Friday (Orthodox)

16 Orthodox Easter Sunday (Orthodox)

21 Eid al-Fitr (Islamic celebration of the end of Ramadan) begins at sighting of crescent moon

23 St George's Day – patron saint of England

23 Start of British asparagus season (ends at summer solstice, 21st June)

APRIL AT A GLANCE

Easter falls this month. It is, of course, a Christian festival, at least the second-most important in the year, but it also has themes that are echoed in the natural world around us, and that some believe came from pre-Christian or pagan roots. The eggs are the main giveaway, the common theme between the Christian story of Easter and the way we actually celebrate it being birth and resurrection. Our Easter baskets are filled with eggs, chicks and baby bunnies, and you only have to spend an hour outdoors at this time of year to understand why. After the long death of winter, young life has returned, with a vengeance. Old gnarled, bare branches are suddenly sprouting fresh little lime-green shoots; bulbs burst from cold, dark soils; lambs are gambolling in the fields; birds sing, court mates, build nests and start laying eggs. Where not long ago the countryside was all angles and jagged branches against the sky, now a fuzz of green covers trees and hedges, and the cherry, crab apple and damson blossom is mimicking the fluffy clouds in the sky. All is now as soft, pretty and pastel-coloured as an Easter bonnet.

Life has returned, and we feel it. There is a rising energy in us that matches that of the outside world – if it hasn't quite reached you, then try to get outside and catch some of it. Sit in your garden or a park on a sunny day and listen to life cranking up all around you. Pick out the sounds of the birds and the bees and of life swinging into full flow. Feel the ever-warming sun on your face and the life flowing back into you.

THE SKY AT NIGHT

Venus – which is always an evening or a morning star because it is closer to the sun than we are – is high in the sky at the end of the month, setting well after sunset. A rare type of eclipse, called a hybrid solar eclipse, occurs this month, though sadly it will not be visible in the UK and Ireland. There is the possibility of a brief glimpse of Mercury in the evening sky.

11th: Mercury at greatest eastern elongation. Mercury is the closest planet to the sun, and this makes it hard to spot, as it is usually lost in the sun's glare. When it is at its furthest point from the sun in our sky, there is a chance to see it for a week or so. Look low in the western sky up to an hour after sunset.

20th: Hybrid solar eclipse. Visible from the region between Australia and Indonesia but not from the UK and Ireland.

22nd–23rd: Lyrids meteor shower. Active 16th–25th April, it peaks late on the 22nd and into the early hours of the 23rd. As we move on our annual orbit around the sun, we pass through the dust trail from Comet Thatcher, which last passed us in 1861, and its debris burns up as it hits our atmosphere. It produces up to 18 meteors per hour, some of them persistent, ionised gas trails that glow for a few seconds. The moon is a thin crescent just past new, so there could be good dark-sky conditions.

23rd: Close approach of the moon and Venus. First visible in the dusk at around 20.30 in the west at an altitude of 33 degrees. They will then go on to set in the northwest at about 23.40.

THE SOLAR SYSTEM

Mercury

Mercury is the smallest planet in the solar system and the nearest to the sun, which makes it tricky for us to see: it is so often lost in the sun's glare. This month, on the 11th, it will be at its greatest eastern elongation, and this can sometimes provide an opportunity to see it in the evening sky, about an hour after sunset.

You may hear astrologers reference the fact that Mercury is in retrograde this month, which means that Mercury changes direction in our skies. This is a remnant from the time when the sun and all of the other planets were thought to orbit around a stationary earth. In fact, of course, all of the planets orbit the sun, but Mercury's orbit laps earth's every four months or so, and when it does it appears to reverse in our skies. If you could take a photograph of the same area of stars every night, Mercury would appear to loop back on itself for about three weeks and then carry on in its usual direction. It will occur between 21st April and 15th May. Astrologers believe that this creates a period of chaos: tempers fray, communication breaks down, technology malfunctions and carefully laid plans go awry.

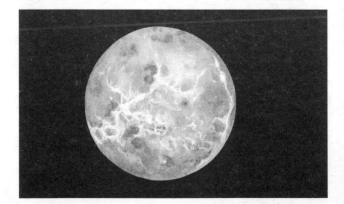

Sunrise and set
Haltwhistle, Northumberland

THE SEA

Average sea temperature in Celcius

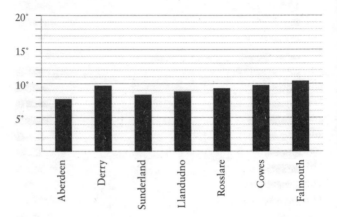

Spring and neap tides

Spring tides are the most extreme tides of the month, with the highest rises and the lowest falls, and they follow a couple of days after the full moon and new moon. These are the times to choose a low tide and go rock-pooling, mudlarking or coastal fossil-hunting. Neap tides are the least extreme, with the smallest movement, and they fall in between the spring tides.

Spring tides: 8th–9th and 21st–22nd

Neap tides: 14th–15th and 28th–29th

Spring tides are shaded in black in the chart opposite.

April tide timetable for Dover

For guidance on how to convert this for your local area, see page 8.

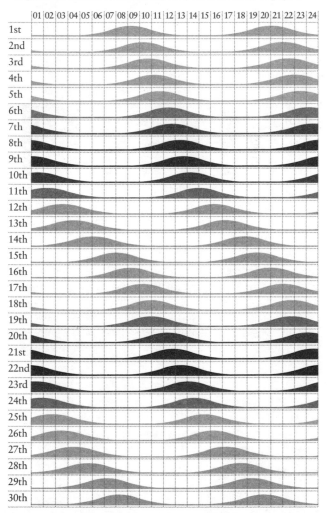

THE MOON

Moon phases

Full moon – 6th April, 05.35*

Last quarter – 13th April, 10.11

New moon – 20th April, 05.12

First quarter – 27th April, 22.19

Moonrise and set

Like the sun, the moon rises roughly in the east and sets roughly in the west. It also rises around 50 minutes later each day. Use the following guide to work out approximate moonrise times.

Full moon: Rises near sunset, opposite the sun, so in the east as the sun sets in the west.
Last quarter: Rises around midnight, and is at its highest point as the sun rises.
New moon: Rises at sunrise, in the same part of the sky as the sun (and so cannot be seen).
First quarter: Rises near noon, and is at its highest point as the sun sets.

Full moon

April's full moon is known as the Budding Moon, New Shoots Moon or Seed Moon. It is also the Paschal Full Moon or the Ecclesiastical Full Moon. Easter is celebrated on the first Sunday after the Paschal Full Moon.

New moon

This month's new moon, on the 20th, is in Aries. Each new moon has its own energy, depending on the zodiacal sign that it is in. The Aries new moon is said to rule creativity and courage.

Moon phases for April

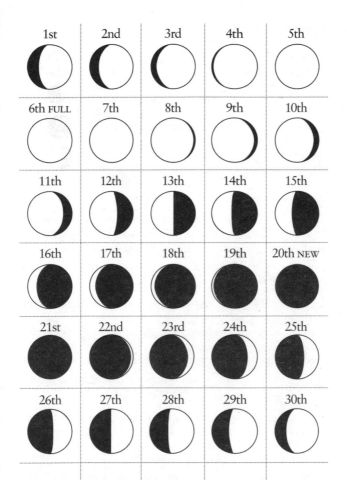

*This full moon falls in the early hours of the morning. To catch it at its fullest during normal waking hours, view it the evening before.

GARDENS

To enjoy this month

Ornamental: Tulips, forget-me-nots, pasque flowers, oxlips, primulas, pulmonaria, scilla, fritillaries, anemones, pussy willow, cherry blossom, apple blossom, pear blossom, cow parsley, bluebells, wood anemones, trillium, wallflowers
Edible: Asparagus, nettle tips, green garlic, sorrel, purple sprouting broccoli, radishes, spring greens, forced rhubarb

Gardening by the moon
The following is a guide to planting with the phases of the moon, according to traditional practices. It also works as a guide to the month's gardening for moon-gardening cynics, who can do these jobs whenever they wish during the month ahead.

First quarter to full moon: 29th March–5th and 28th–5th May
This is the best time for sowing crops that develop above ground, but is bad for root crops. Plant out seedlings and young plants. Take cuttings and make grafts. Avoid any other pruning. Fertilise.
- In pots under cover sow French beans, runner beans, cabbages, cauliflowers, courgettes, cucumbers, Florence fennel, kale, pumpkins and winter squashes, sweetcorn.
- Sow herb seed – coriander, chervil, dill, basil and parsley in pots or seed trays indoors, or into the soil under cloches.
- Sow lettuces, peas, broad beans, rocket, summer purslane, corn salad, spinach and Swiss chard direct.
- Sow hardy flower seed in pots or direct.
- Plant up pots and hanging baskets with bedding plants but keep them under cover until the danger of frost has passed.

Full moon to last quarter: 6th–13th (till 10.11)
A 'drawing down' energy. This phase is a good time for sowing and planting any crops that develop below ground: root crops, bulbs and perennials. Light is high but decreasing.
- Plant second early and maincrop potatoes.
- Sow direct in light soils in warm areas, covering with

cloches after: carrots, beetroot, parsnips, turnips, leeks, spring onions.
- Plant out asparagus crowns and globe artichokes, grapevines and strawberries.
- Plant lilies and gladioli.

Last quarter to new moon: 13th (from 10.11)–19th

A dormant period, with low sap and poor growth. Do not sow or plant. A good time though for pruning, while sap is slowed. Weeding now will check growth well. Harvest any crops for storage. Fertilise and mulch the soil. Garden maintenance.
- Earth up first early potatoes.
- Harvest asparagus, lettuces, rocket and winter salad leaves, spring onions, rhubarb, and the last of the purple sprouting broccoli, leeks and kale.
- Start to harden off plants that have been grown outdoors, moving them outside during the day and back in at night.
- Push shrubby peasticks into the ground to support peas.
- Build bamboo supports for French and runner beans.
- Weed and cut the lawn.

New moon to first quarter: 20th–27th

The waxing of the moon is associated with rising vitality and upward growth. Towards the end of this phase plant and sow anything that develops above ground. Prepare for growth.
- Sow any of the seeds in the first quarter–full moon phase towards the end of this period.
- Pot houseplants on into the next plant pot size, watering them regularly and feeding them fortnightly
- Prepare ground, mulch and feed soil.
- Remove covers from forced rhubarb. Feed perennials with liquid feed. Feed and mulch roses.

Note: Where no specific time for the change between phases is mentioned, this is because it happens outside of sensible gardening hours. For exact changeover times, refer to the moon phase chart on page 87.

THE RECIPES

Bun of the month

Marzipan hot cross buns

In Britain and Ireland at least, hot cross buns are the most well known and regularly eaten of all celebration buns. Their history stretches right back to 1361, when Brother Thomas Rodcliffe of St Albans Abbey began baking them to distribute to the poor on Good Friday. They were particularly popular in the 18th century and have been eaten at Easter ever since. They should traditionally be baked and eaten on Good Friday, though hot cross bun season in the shops now can be all year round, and they are hard to resist. Those baked on Good Friday do have special properties, though. They will not go mouldy, and one hung in your kitchen will ensure that all of your bread turns out perfectly in the year ahead. They protect against fires and shipwrecks, too. All very good reasons to bake your own. The one less delicious part of traditional hot cross buns is the flour-paste cross, and so in this version the cross is formed from marzipan.

Makes 12
Ingredients
300ml milk, plus extra for the eggwash
50g butter
500g strong white bread flour, plus extra for dusting
1 sachet (7g) instant yeast
75g brown sugar
1 teaspoon mixed spice
1 teaspoon cinnamon
1 teaspoon salt
3 eggs
150g sultanas
80g mixed peel

Zest of 1 orange

Zest of 1 lemon, juice of ½

140g icing sugar, sieved

75g caster sugar

200g ground almonds

2 tablepoons apricot jam

A

Method

Line a baking tray with parchment. Gently heat the milk
and butter in a saucepan until melted, then allow to cool.
Combine the flour, yeast, brown sugar, mixed spice, cinnamon
and salt in a bowl. Make a well in the middle, crack 1 egg into
it, add the cooled milk and combine to form a sticky dough.
Add the sultanas, mixed peel and the orange and lemon zest;
combine. Turn onto a floured surface and knead for about 5
minutes. Pop the dough back in the bowl, cover and leave in a
warm place for about an hour, to double in size.

While the dough is rising, make a small batch of marzipan
by mixing the icing sugar, caster sugar, ground almonds,
1 egg and the lemon juice, and bring together into a ball. If it
is too sticky, add more icing sugar. Chill in the refrigerator for
30 minutes.

Knock the air out of the dough and form it into 12 round
balls, placing them on the baking tray. Cover and allow to rise
for another hour.

Meanwhile, preheat the oven to 190°C, Gas Mark 5. Cut
off a third of the marzipan and roll it into a rectangle a few
millimetres thick. When the buns have risen, beat 1 egg with a
splash of milk to form an egg wash, and brush over the buns.
Cut 2 long strips of the marzipan and lay them across each
bun to form a cross, making sure they are firmly attached.
Bake for 20 minutes, keeping an eye on the marzipan to check
it doesn't burn. When the buns are done, heat the apricot jam
and brush it over the buns while they are still warm.

Torta pasqualina

This traditional Easter pie of greens, eggs and cheese is originally from Liguria in northwest Italy and makes a good vegetarian alternative for Easter Day.

Serves 6
Ingredients
1 packet (500g) all butter puff pastry
3 tablespoons olive oil
1 onion, chopped
1kg chard, stems removed and chopped and leaves sliced
9 eggs
350g ricotta cheese
85g Parmesan cheese, finely grated
Handful of dill, chopped
Nutmeg
Salt and pepper

Method
Line a 20-cm springform tin with parchment. Cut off a third of the pastry and set aside. Roll the rest out on a floured surface and line the tin with it, pressing the pastry into the corners and trimming off any excess. Roll out the reserved third to make a lid.

Heat the oil in a large saucepan, add the onion and cook for about 10 minutes or until it begins to soften and turn translucent. Add the chard stems and cook for a further 5 minutes. Then add the leafy parts of the chard and cook for a further 5 minutes. Tip all onto a clean tea towel and pull up the corners of the tea towel so that all of the greens mixture falls into the centre. Spin the ball of greens and squeeze out as much moisture as you can, making it as dry as possible.

Preheat the oven to 190°C, Gas Mark 5. In a large bowl whisk together 3 eggs then set aside 2 tablespoons of it. Stir the ricotta, half the Parmesan, the dill, some grated nutmeg, the greens and salt and pepper into the beaten eggs. Tip the mixture into the pastry case, make 6 indents with a tablespoon and crack an egg into each. Sprinkle with the remaining Parmesan and fit the lid on top, crimping the edges together to seal, and brushing with the reserved beaten egg. Pierce the lid a few times with a sharp knife. Bake for 50 minutes, or until golden.

THE ZODIAC

Aries: 21st March–19th April

The sun begins the month in the same area of sky that holds the constellation of Aries, the Ram, the 0–30th degree of the zodiac. On the 20th of this month the sun will move into Taurus (see page 116).

> **Symbol:** The Ram
> **Planet:** Mars
> **Element:** Fire
> **Colour:** Red
> **Characteristics:** Strong, loyal, blunt, fiery, impulsive, energetic

A Greek king, Athamas, took the goddess Nephele as his wife, and she gave birth to two children: a boy, Phrixus, and a girl, Helle. Nephele abandoned the king, and a great drought took hold of the land, which the king's second wife, Ino, decided would be broken only by sacrificing Phrixus. Nephele appeared to her children with a winged golden ram, and they climbed onto its back and were carried across the sea to Colchis, in modern-day Georgia. Phrixus sacrificed the ram to Zeus and hung its golden fleece in a grove sacred to Aries, the god of war. The ram became the constellation Aries, and the fleece came to represent authority and kingship. It was later stolen by Jason and the Argonauts to help Jason gain his rightful place on the throne of Iolcus in Thessaly. The best time to spot Aries is when it is in the opposite part of the sky from the sun six months from now, in October.

A

A FOLK SONG FOR ARIES' RAM

'The Derby Ram'
Traditional, arr. Richard Barnard

This ram has its monstrous size in common with the herring in March's song (see page 74). It easily rivals the magical properties of the ram of the golden fleece myth, which is immortalised in the sky as Aries, this month's zodiac sign.

The song is thought, like so many English folk songs, to originate in a midwinter mummers' custom: one man would dress as the ram and be accompanied by a group of men, singing and raising money for beer. This died out last in the English Midlands, from where this song originates.

As I was going to Der - by all on a mar - ket day, I met the fi - nest ram, sir, that e - ver was fed on hay. In - deed, sir, it's true, sir, I ne - ver was one to lie, and if you had been to Der - by, you'd have seen him as well as I.

As I was going to Derby, all on a market day,
I met the finest ram, sir, that ever was fed on hay.
Indeed, sir, it's true, sir, I never was one to lie,
And if you had been in Derby, you'd have seen him as well
as I.

He had four feet to walk upon, he had four feet to stand,
And every foot that he put down, it covered an acre of land.
Indeed, sir, it's true, sir, I never was one to lie,
And if you had been in Derby, you'd have seen him as well
as I.

The horns upon his head, sir, they reached up to the sky,
And on them was an eagle's nest: I heard the young ones cry.
Indeed, sir, it's true, sir, I never was one to lie,
And if you had been in Derby, you'd have seen him as well
as I.

The man that fed the ram, sir, he fed him twice a day,
And every time he fed the ram, he ate a rick of hay.
Indeed, sir, it's true, sir, I never was one to lie,
And if you had been in Derby, you'd have seen him as well
as I.

The wool that grew on his sides, sir, made fifty packs
complete,
And that was sent to Flanders to clothe the British fleet.
Indeed, sir, it's true, sir, I never was one to lie,
And if you had been in Derby, you'd have seen him as well
as I.

The tail was fifty yards, sir, as far as I could tell,
And that was sent to Rome, sir, to ring St Peter's bell.
Indeed, sir, it's true, sir, I never was one to lie,
And if you had been in Derby, you'd have seen him as well
as I.

NATURE

The pond in April

There is an explosion of green here in April, as frogbit, hornwort, water soldiers and broad-leaved pondweed rise up from the mud at the bottom, reach the surface and spread their leaves, ready to make the most of the warmth and daylight. Water lilies do the same, spreading their pads across the surface, providing little islands for any creatures needing a rest from all of their activity.

And there are plenty. We are into the main breeding season now, and every creature in the garden has a use to make of the pond and its surroundings. Swallows and house martins collect mud from the edges to build their nests. Moths, leaf miners and aphids lay their eggs on the leaves around the pond. Sparrows swoop in and take the aphids and caterpillars for their own babies, and bathe and drink at the edges. Yellow flag iris and marsh marigold are in flower, and the bees visit them for pollen and nectar. When they get weary, they might rest on a lily pad and take a sip of pond water before continuing on with their work. The hatched tadpoles have now started to grow legs, and they switch from being vegetarian to eating dead insects from the water's surface. This is a great time for pond-dipping. A bucketful of water will be alive with water boatmen, pond snails and dragonfly larvae.

There are three newts native to the UK – the smooth newt (by far the most common), the palmate newt and the great crested newt – and one to Ireland – the smooth newt. Newts are secretive breeders, choosing deeper ponds, and carrying out most of their breeding by night; shine a torch into your night-time pond to try to spot one. They begin breeding some time in April and continue for up to six weeks. In the run-up to breeding, the male puts on a show, brightening his colours and extending the crest that runs along his back, and then performing a ritualistic dance, swishing, fanning and whipping his tail. When the female accepts him, he will touch her with his tail to transfer sperm. The female wraps each individual fertilised egg in a leaf, laying up to 600 between now and July.

A

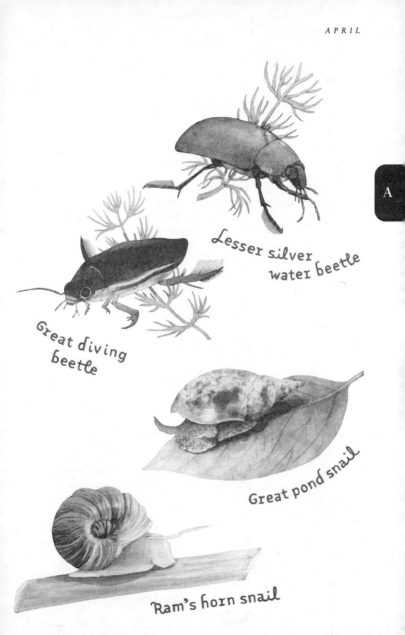

Lesser silver
water beetle

Great diving
beetle

Great pond snail

Ram's horn snail

May

 May Day (traditional)

 Beltane (Gaelic/Pagan May Day celebration)

 Early May bank holiday, England, Wales, Scotland, Northern Ireland. May Day bank holiday, Ireland

1 International Workers' Day

7 International Dawn Chorus Day

11 Stow Horse Fair – Gypsy and Traveller spring gathering

14 Rogation Sunday/beating the bounds (Christian/traditional)

18 Ascension Day/Holy Thursday (Christian)

25 Shavuot/Feast of Weeks begins at sundown (Jewish)

28 Pentecost/Whit Sunday (Christian)

29 Spring bank holiday, England, Wales, Scotland, Northern Ireland

MAY AT A GLANCE

There are certain days in the year that just have magic about them, a shimmer in the air. May Day, or Beltane, is one of these. Perhaps it is the ghostly trace of so many celebrations carried out over hundreds of years that echoes and calls us to mark this day: fêtes and processions, May queens and Jacks-in-the-green, Morris dancers and maypoles. Further back in time, young women ran through meadows to splash their faces with the dew before dawn, young men leapt over bonfires, and cattle were dressed in flowers and ceremoniously driven from their winter quarters and out to pasture. This has always been a bit of a hedonistic moment, with a theme of liberation – the final throwing off of the ties and confines of winter. It's a moment to head into summer and the freedoms it promises, face daubed green and flowers woven wildly into hair. The natural world is certainly at it. Every flower is trying to attract a bee, and every bird to find its mate. Behind it all is an urgency: there are only so many warm months in the year to get your offspring born, grown and toughened up for winter – be they baby dormice, sparrows or seeds – so let's go.

If the countryside looks like it's having a party, that's because it is. The hedgerows are foamy with hawthorn and cow parsley and dotted with fat golden sunflowers and bluebells, soundtracked by the buzzing of bees and the glorious racket of the dawn chorus. This is a beautiful moment in the year, a moment to make merry and dance, for the casting of clouts and the having of fun. Allow yourself to feel the sense of release that comes with warmer weather, with light and flowers. This cusp of spring and summer is stunning, and worth the celebration.

THE SKY AT NIGHT

This is the best month of this year for spotting Venus, which will be getting steadily more prominent, setting some four hours after sunset by the middle of the month. After that, it will get closer to the sun and gradually less visible again, and will be lost in the glare of our sunset by mid-July.

5th: Penumbral lunar eclipse, not visible from the UK or Ireland. It will be a slight dimming as the moon passes through the earth's penumbra, or partial shadow. It will be visible through Asia and Australia, and some of Eastern Europe and eastern Africa.

6th–7th: Eta Aquarids meteor shower. This meteor shower occurs when we pass annually through the dust trail left by Halley's Comet, the debris burning up as it hits our atmosphere. In the northern hemisphere we will see only around 30 meteors per hour at the peak in the early hours of the 7th, but in the southern hemisphere there could be up to 60 per hour. We will be just past the full moon, which will mean we will miss all but the brighter trails.

13th: Close approach of the moon and a dim Saturn. They rise at around 03.30 in the east and become lost in the dawn in the southeast at about 04.40 at an altitude of 11 degrees above the horizon.

23rd: Close approach of the moon and Venus. They are first visible in the dusk at around 21.20 in the west at an altitude of 30 degrees above the horizon. They then go on to set at around 24.00 in the northwest.

29th: Mercury at greatest western elongation. Mercury is the closest planet to the sun, which makes it hard to spot, as it is usually lost in the sun's glare. Towards the end of the month it will be the furthest it gets away from the sun in our sky, and so you may spot it low in the eastern sky a little before sunrise.

M

THE SOLAR SYSTEM

Venus

Venus is the second planet from the sun and very much visible to the naked eye from earth: after the moon it is the brightest object in our night sky. It is often called the 'morning star' or the 'evening star' because it travels around the sun within earth's orbit. It is therefore in the same part of our sky as the sun, either rising shortly before it and lost in its glare at dawn, or following shortly behind it, sinking below the horizon not long after dusk. Currently it is an evening star and is at its furthest away from the sun and most visible, setting several hours after sunset.

Like earth, Venus is a rocky planet and it is similar to earth in size. There may once have been water on its surface but this long ago vanished – Venus has a surface temperature of 464°C, and it is now covered in a dense cloud of sulphuric acid.

Venus is named after the Roman goddess of love and is associated with this month's star sign, Taurus, over which it is said to rule.

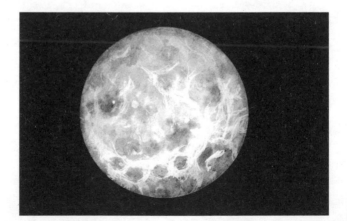

Sunrise and set
Haltwhistle, Northumberland

THE SEA

Average sea temperature in Celcius

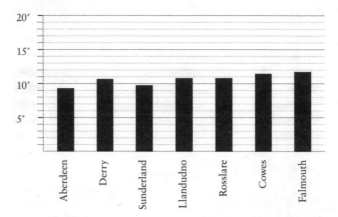

Spring and neap tides

Spring tides are the most extreme tides of the month, with the highest rises and the lowest falls, and they follow a couple of days after the full moon and new moon. These are the times to choose a low tide and go rock-pooling, mudlarking or coastal fossil-hunting. Neap tides are the least extreme, with the smallest movement, and they fall in between the spring tides.

Spring tides: 6th–7th and 20th–21st

Neap tides: 13th–14th and 28th–29th

Spring tides are shaded in black in the chart opposite.

May tide timetable for Dover

For guidance on how to convert this for your local area, see page 8.

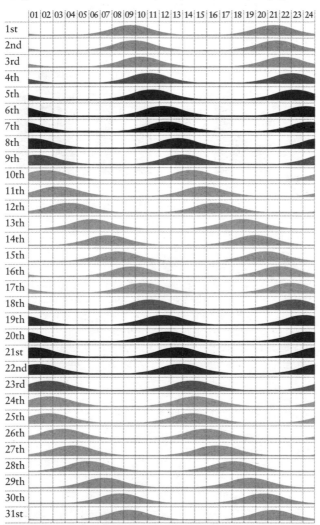

THE MOON

Moon phases

Full moon – 5th May, 18.34

Last quarter – 12th May, 15.28

New moon – 19th May, 16.53

First quarter – 27th May, 16.22

Moonrise and set

Like the sun, the moon rises roughly in the east and sets roughly in the west. It also rises around 50 minutes later each day. Use the following guide to work out approximate moonrise times.

Full moon: Rises near sunset, opposite the sun, so in the east as the sun sets in the west.
Last quarter: Rises around midnight, and is at its highest point as the sun rises.
New moon: Rises at sunrise, in the same part of the sky as the sun (and so cannot be seen).
First quarter: Rises near noon, and is at its highest point as the sun sets.

Full moon

May's full moon is known as the Mother's Moon or Bright Moon.

New moon

This month's new moon, on the 19th, is in Taurus. Astrologers believe that the new moon is a quiet, contemplative time before a phase of growth. Each new moon has its own energy, depending on the zodiacal sign that it is in. The Taurus new moon is said to rule practicality, sensuality and physicality.

Moon phases for May

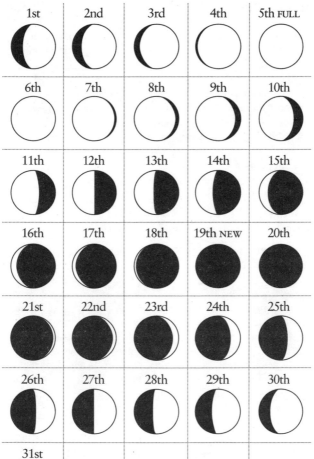

GARDENS

To enjoy this month

Ornamental: Bluebells, tulips, lilac, ornamental alliums, foxgloves, irises, hawthorn blossom, sweet woodruff, lily-of-the-valley, apple blossom, cow parsley, elderflowers, blue alkanet, camassia, Canterbury bells, sweet William, honesty, sweet rocket, Iceland poppies, dandelions, marigolds
Edible: Asparagus, broad beans, gooseberries, peas, garlic, lettuces, spring onions, radishes, forced rhubarb, elderflowers, sorrel, watercress, chives and chive flowers, borage, chamomile

Gardening by the moon
The following is a guide to planting with the phases of the moon, according to traditional practices. It also works as a guide to the month's gardening for moon-gardening cynics, who can do these jobs whenever they wish during the month ahead.

First quarter to full moon: 28th April–5th (till 18.34) and 27th (from 16.22)–3rd June
This is the best time for sowing crops that develop above ground, but is bad for root crops. Plant out seedlings and young plants. Take cuttings and make grafts. Avoid any other pruning. Fertilise.

- Direct sow corn salad, French beans, runner beans, sweetcorn, Brussels sprouts, calabrese, cauliflowers, Florence fennel, kale, lettuces, peas, spinach and purple sprouting broccoli, plus cucumbers under cloches. Continue sowing small amounts of herb seed.
- Sow Brussels sprouts and broccoli. Start cucumbers, courgettes, pumpkins and squashes in pots indoors.
- Plant out vegetable and bedding plants and hanging baskets towards the end of the month, or when you are confident that frosts have passed.

Full moon to last quarter: 5th (from 18.34)–12th (till 15.28)
A 'drawing down' energy. This phase is a good time for sowing
and planting any crops that develop below ground: root crops,
bulbs and perennials. Light is high but decreasing.
- Lift, split and replant daffodil clumps to revitalise them.
  Split and replant spring-flowering perennials.
- Direct sow beetroot, carrots, kohlrabi, radishes, spring
  onions, swedes and turnips.
- Plant lilies, dahlias, gladioli.

M

Last quarter to new moon: 12th (from 15.28)–19th (till 16.53)
A dormant period, with low sap and poor growth. Do not sow
or plant. A good time though for pruning, while sap is slowed.
Weeding now will check growth well. Harvest any crops for
storage. Fertilise and mulch the soil. Garden maintenance.
- Earth up potatoes.
- Prune spring-flowering clematis.
- Tie in climbers and ramblers.
- Liquid feed spring bulbs that have finished flowering.
- Put supports in place for herbaceous perennials.
- Tie in sweet peas as they grow.
- Net fruit bushes to protect them from the birds.

New moon to first quarter: 19th (from 16.53)–27th (till 16.22)
The waxing of the moon is associated with rising vitality and
upward growth. Towards the end of this phase plant and sow
anything that develops above ground. Prepare for growth.
- Sow small amounts of herb seed in pots or seed trays
  indoors, or direct into the soil under cloches.
- Pot chillies, tomatoes and any other vegetable plants that
  you are growing in containers into their final pots.

Note: Where no specific time for the change between phases
is mentioned, this is because it happens outside of sensible
gardening hours. For exact changeover times, refer to the moon
phase chart on page 109.

THE RECIPES

Bun of the month

Herb and ricotta Ukrainian *pyrizhky*

Pyrizhky are small, savoury buns eaten throughout Ukraine and usually stuffed with cabbage or meat. This seasonal ricotta and herb version is adapted, with kind permission, from Olia Hercules' recipe in her book *Mamushka*, and is full of fresh late-spring flavours. Olia is one of the co-founders of Cook for Ukraine, a global network of chefs and cooks raising money for UNICEF's Ukraine appeal. Find out more at www.cookforukraine.org.

Makes 8

Ingredients

250ml milk

100ml olive oil, plus extra for frying

1 teaspoon caster sugar

300g strong white bread flour, plus extra for dusting

100g spelt flour

1 sachet (7g) instant yeast

1 tablespoon salt

1 tablespoon ground turmeric

For the filling:

200g ricotta or curd cheese

20g dill, chopped

20g chives

1 red onion, finely sliced

1 garlic clove, finely grated

Zest of 1 lemon

Salt and pepper

Method

Gently warm the milk, olive oil and caster sugar in a
saucepan. In a large bowl, combine the bread flour, spelt flour,
yeast, salt and ground turmeric. Make a well in the centre and
mix the liquid into the dry ingredients until you've made a
soft dough. Now gently knead it until smooth, adding a little
flour to help if it's too sticky, but don't add too much flour
or it won't be as light. Cover and leave somewhere warm for
about 1 hour, until it has doubled in size.

Make the filling by mixing together the ricotta or curd
cheese, dill, chives, red onion, garlic and lemon zest. Season
with plenty of salt and pepper.

Knock the air out of the dough and separate into 8 balls.
On a floured surface, roll out each to the size of a saucer,
making sure the dough isn't too thick, as it needs to cook
through in a short time. Place a heaped tablespoon of the
filling on each, then take up opposite sides of the circle and
pinch along the top to seal. Heat a few tablespoons of olive
oil in a frying pan over a medium heat, and place half the
buns in the hot oil, seam down. Fry for 3–4 minutes on each
side until golden. Dab off any excess fat on kitchen paper and
repeat with the remaining buns (be careful not to over-crowd
the pan and fry in smaller batches if necessary). Serve as soon
as the *pyrizhky* are cool enough to eat.

A cake for St Honoré's Day

On hearing that the sixth-century French nobleman Honoratus (later to become St Honoré) had been made Bishop of Amiens, his old nursemaid – who was baking bread for the family – said that she would only believe it if the peel she used to remove the bread from the oven took root and grew. Of course it did, and it grew into a mulberry tree that lived for several hundred years. Perhaps this story is why Honoré became patron saint of pastry chefs, and the very fancy Gâteau St-Honoré is made in his honour on his feast day, 16th May. It is designed for showing off, using lots of different techniques. This is a simplified version, though still fancy enough to mark the day. It makes a very summery celebration cake with its pretty strewing of early summer fruits and a few edible flowers from the garden.

Serves 6–8

Ingredients

For the choux pastry:

100ml milk

150ml water

100g butter, cubed

1 tablespoon caster sugar

175g plain flour

4 large eggs

For the caramel:

100g caster sugar

For the filling:

150ml vanilla custard

250ml double cream

50ml gin

1 teaspoon lemon extract

1 teaspoon ground cardamom

For the topping:

6 fresh apricots, sliced into eighths, and/or a handful of quartered strawberries

30g pistachio kernels, lightly toasted and roughly chopped

A few edible flowers, such as primroses, violets, elderflower, lilac or marigold petals (optional)

Method

Preheat the oven to 220°C, Gas Mark 7 and line 2 large baking trays with parchment. Gently heat the milk, measured water, butter and sugar in a saucepan until the butter has melted. Add the flour all at once and beat with a wooden spoon over the heat for about 1 minute, bringing it into a mushy ball in the pan. Remove from the heat and continue to beat for another minute. Stir in the eggs one by one, beating between each addition until you have a smooth and velvety choux paste. Place into a piping bag and pipe a spiralling circle the size of a dinner plate on one of the baking trays. On the other tray, use the remaining choux paste to pipe small balls, using a wet finger to smooth any pointy tops. Bake the pastries for 10 minutes, then turn the oven down to 200°C, Gas Mark 6 and bake for a further 10–15 minutes, until nicely browned and well risen. Remove from the oven and carefully pierce each bun with a sharp knife to let out any extra steam. Pierce the round base too before transferring the pastries to a wire rack.

When the buns have cooled, make the caramel. Heat the sugar in a small frying pan over a high heat. Do not stir the mixture but gently swirl to allow all the sugar to melt, making sure it doesn't burn. When it becomes a dark golden liquid, remove from the heat. Carefully dip the bottom of each bun in the caramel, placing it back on the cooling rack bottom-side-up and allow it to set. Repeat for all the buns, reheating the caramel if it becomes solid. Make patterns on greaseproof paper for decoration with any leftover caramel.

To make the filling, whisk the custard and cream until soft peaks form. Add the gin, lemon extract and cardamom and whisk again until stiff peaks form. Fit a piping bag with a small nozzle and fill it with half the mixture. Make a little hole in the side of each bun and pipe in the custard cream until plump. Place the large base on a pretty plate and spread the rest of the cream across it. Arrange the choux buns around the rim of the circle, then fill the centre with the fruits, scattering the pistachios and edible flowers over the top.

THE ZODIAC

Taurus: 20th April–20th May

The sun begins the month in the same area of sky that holds the constellation of Taurus, the Bull, the 30th–60th degree of the zodiac. On the 21st of this month the sun will move into Gemini (see page 139).

> **Symbol:** The Bull
> **Planet:** Venus
> **Element:** Earth
> **Colour:** Pink
> **Characteristics:** Honest, stubborn, trustworthy, sensual, passionate, ambitious

Taurus is one of the oldest named constellations, picked out of the sky by the Mesopotamians some time between 4000 BCE and 1700 BCE. The Babylonians, Greeks, Romans and ancient Egyptians all connected this constellation with a bull. But, of course, the ancient Greeks had a story for it. Yet again it involves a less than charming tale of Zeus and his libido. Zeus had a habit of transforming himself into the shape of other creatures in order to 'seduce' – a euphemism – various goddesses and mortals. He transformed himself into a tame white bull to attract the attention of the Phoenician princess Europa. She stroked his flanks and then sat on his back, at which point he rose and ran with her into the sea, then swam to Crete. They had three sons, one of whom was Minos, who went on to become the king of Crete (and who had the labyrinth built to contain the Minotaur). To commemorate this, Zeus later put the bull into the sky as a constellation. The best time to spot Taurus is when it is in the opposite part of the sky from the sun six months from now, in November.

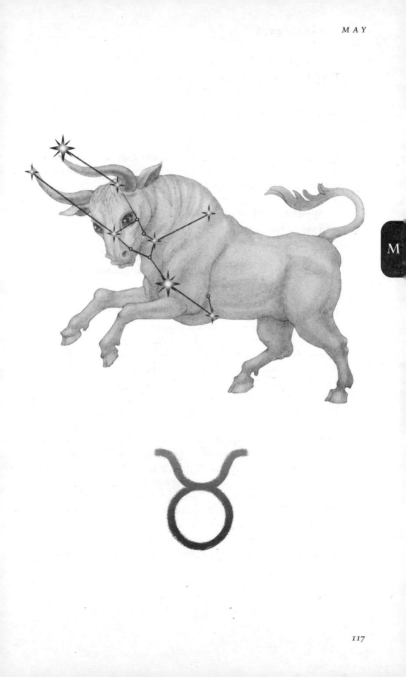

A FOLK SONG FOR TAURUS' BULL

'How I Could Ride!'
Words traditional, music by Richard Barnard

A 'chorus song', short and simple with a repeated refrain, and a bit silly, as so many animal songs seem to be, is used here to represent Taurus' bull. The words were collected by Alfred Williams as part of his pre-World War I mission to document folk songs from Wiltshire and surrounding areas before they disappeared. He didn't record the tunes, though, so composer Richard Barnard has written this one to fit the words.

How I could ride if I had but a horse!
My mother cried, 'Take the bull in the close.'
So I took the old bull and he cocked up his tail
And I went away in a storm of hail.

How I could ride if I had but a bridle!
My mother cried, 'Don't you be so idle!'
So I took the old bull and he cocked up his tail
And I went away in a storm of hail.

How I could ride if I had but a saddle!
My mother cried, 'Take the rags from the cradle.'
So I took the old bull and he cocked up his tail
And I went away in a storm of hail.

How I could ride if I had some spurs!
My mother cried, 'Take the nails from the doors.'
So I took the old bull and he cocked up his tail
And I went away in a storm of hail.

M

NATURE

The pond in May

The pond plants are at their lushest and freshest now and putting on a great burst of growth, quickly threatening to cover the pond. Many are in flower and buzzing with bees. Cuckoo flower, also known as lady's smock, is producing its pale pink flowers, as is water crowfoot – producing its white flowers just above the surface of the water – and the closely related and golden-coloured water buttercup.

Look out for the magical transformation that occurs when a dragonfly nymph hauls itself out of the water. They often choose vertical reeds or the leaves of flag irises, and there the nymph will shed its final exoskeleton, having moulted up to 17 times in the process of growing from egg to adult. At the moment that it climbs out of the top of the exoskeleton, it is the same drab brown as it has been all through its life. It will rest for one to three hours, allowing its body and wings to harden up, and will then take a maiden flight of a few metres. Over the following days and weeks, it will become stronger and take on its vibrant adult colouring, ready to mate.

Tadpoles are beginning to resemble something close to tiny frogs now, while newt eggs are starting to hatch, into larvae rather than tadpoles. They have gills and will live underwater for now, growing into newts over around 90 days. After mating, the older newts leave the pond and head back to the territory they have established on land.

The water now is alive with activity. Dytiscidae beetles dive and hunt among the fern-like spiked water milfoil and the fennel pondweed, larvae wriggle and dart about below the surface, and water boatmen skate across the surface.

Whirligig beetle

Backswimmer

M

Lesser water
boatman

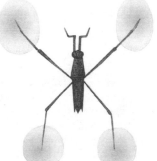

Common pond
skater

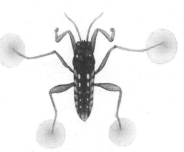

Water cricket

June

 Start of meteorological summer

 Start of Pride month

 Start of Gypsy, Roma and Traveller History Month

 Trinity Sunday (Christian)

 June bank holiday, Ireland

 Corpus Christi (Christian)

 8th–11th: Appleby Horse Fair – Gypsy and Traveller gathering

 17th–25th: Special Olympics World Games, Berlin

 Father's Day

 Summer solstice, at 15.57/midsummer – start of astronomical summer

 Litha (Pagan midsummer celebration)

 World Humanist Day

 St John's Day/traditional Midsummer Day/ Feast of St John the Baptist (Christian/ traditional)

 26th–1st July: Hajj – Islamic pilgrimage to Mecca

JUNE AT A GLANCE

Another month containing another enchanted time: Midsummer Eve and Midsummer Day. There is only a smattering of such days through the year, but they come thick and fast in early summer. Maybe it's the quality of the light that encourages fairies, filtering and pooling through the bright or deep green of midsummer foliage. Or perhaps it's the long, languorous evenings, barely turning into true night at all, that have us wanting to stay out late, light fires that spark against the deep blue summer gloaming, and feel the magic of the year.

Midsummer was traditionally celebrated on 24th June, St John's Day – he was the man who foretold Jesus' birth. Unusually among feast days, it celebrates St John the Baptist's birth rather than his death, and it has a particular significance because it falls exactly six months before Jesus' birth. There is always this hint of a mirroring of midwinter at midsummer. We celebrate the joy of light and warmth, knowing that this is the tipping point, when we begin the slow slide towards the cold and dark. And because of this, there is also a strong tradition of attempting to capture the moment and prolong it. Midsummer bonfires were traditionally lit, on beacon hills and in towns and villages, and were kept burning all night. This was one of the few times that long-tended hearth fires would be extinguished, only to be relit using the flame of the midsummer fire. The hearth fires would then be kept burning into the months to come, holding back the coming of winter.

Of course, we can't hold winter back, and we can't capture midsummer – it is as fleeting as all of the other moments of the year. But we can make sure we soak it in. Sit in it, feel the warm breeze against your skin, and make yourself a picture of your surroundings – how blue the sky is and how warm the breeze is. Hold this midsummer feeling within you, and tend it like that flame in the hearth, for as long as you can.

THE SKY AT NIGHT

Saturn becomes visible as a morning star towards the end of the month, rising in the southeast at around 02.00. It will be dim at first but will brighten in the following weeks, building in brightness and visibility towards August, when it will be at opposition (see page 172), and so at its closest and brightest.

10th: Close approach of the moon and Saturn. They rise at around 01.40 in the east. They become lost in the dawn at about 04.00 in the southeast at an altitude of 21 degrees from the horizon.

14th: There will be a brief view of a close approach of the moon and Jupiter. They rise around 03.00 in the east. They become lost in the dawn at about 04.00 in the east at an altitude of 13 degrees above the horizon.

21st: Close approach of the moon and Venus. First visible in the dusk at around 21.50 in the west at an altitude of 18 degrees above the horizon. They then go on to set at about 23.30 in the northwest.

J

THE SUMMER SOLSTICE

We reach the point this month when the axial tilt of the earth is aligned with the sun, with the north pole tipped towards it and the south pole tipped away. Day is much longer than night now in the northern hemisphere, and night is much longer than day in the southern hemisphere. This is the summer solstice in the northern hemisphere and the winter solstice in the southern hemisphere, or more diplomatically, the June solstice.

A couple of the 'circles of latitude' – the five invisible lines that map-makers draw around the globe – become useful at this moment. Of primary importance to the June solstice is the Tropic of Cancer, which runs at a latitude of 23.43 degrees north of the equator and travels through Algeria, Egypt, Saudi Arabia, Bangladesh, China, across the Pacific Ocean, through Mexico and the Bahamas and then across the Atlantic Ocean to northern Africa. This is the most northerly point on the earth where the sun can be directly overhead, and it does this only at the exact moment of the summer solstice, before heading back south again.

Also having its moment now is the Arctic Circle, which runs 66.56 degrees north of the equator (which is equivalent to the latitude of the north pole minus the earth's axial tilt of 23.4 degrees). The Arctic Circle is the southernmost point at which the sun can be above the horizon for 24 hours, and on the Arctic Circle itself this occurs only once a year, on the summer solstice. The further north you travel within the Arctic Circle, of course, the more days there are of midnight sun.

Even south of the Arctic Circle, nights will be very light now, as the sun is never very far below the horizon. This phenomenon, of light spilling over the top of the north pole in summer, explains why summer days are longer in Scotland than they are in the south of England.

Naming of the summer solstice
The word 'solstice' is derived from the Latin *solstitium* – from *sol* ('sun') and *sto* ('stand'), roughly meaning 'the sun stands still'. It refers to the fact that the sun reaches its most

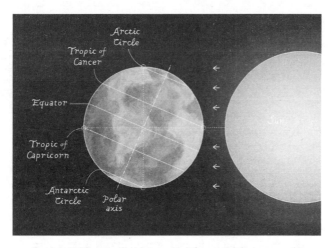

northerly position in the sky as seen from earth, pauses there, and then begins its journey back in the opposite direction. The Welsh name for the summer solstice is *alban hefin* – *alban* meaning 'quarter year' and *hefin* meaning 'summer'. In Scottish and Irish Gaelic it is called *grianstad an tsamhraidh*. Modern Wiccans, Druids and Pagans call this solstice Litha, from an Anglo-Saxon word for June, Ærra-Liða, meaning 'the first Liða' (July is the second). The word also meant 'calm', or 'gentle', perhaps referring to the weather.

Summer solstice facts
Date: 21st June
Time: 15.57
Altitude: At solar noon on the 21st June, the sun will reach 62 degrees in the London sky, 58 degrees in the Glasgow sky and 60 degrees in the Dublin sky.
Sunrise times: London 04.43, Glasgow 04.31, Dublin 04.56
Sunset times: London 21.21, Glasgow 22.06, Dublin 21.56
Sun's distance from earth: 152,028,000 kilometres

Mark the summer solstice

- Cut sprigs of herbs that dry well, such as oregano, thyme, mint, marjoram, rosemary and bay, and hang them up in bunches somewhere airy to dry, and for you to use throughout winter. Herbs were traditionally gathered in this way at midsummer, when they are at their best (ideally at dawn, but that's pretty early at the moment).
- Collect flowers to make into posies to give to friends or to have around your house in jam jars. Include roses, lavender, oregano, cornflowers, herb Robert, honeysuckle, red campion, oxeye daisy and whatever you can find. Tie with colourful ribbons.
- To a spray bottle of water, add 1 teaspoon of vodka and 25–30 drops of summery essential oils (try lavender, ylang ylang and geranium). Shake it up and spritz it around.
- Hang prisms on your sunniest windows to capture the solstice sunlight and turn it into rainbows.

Sunrise and set
Haltwhistle, Northumberland

	01	02	03	04	05	06	07	08	09	10	11	12	13	14	15	16	17	18	19	20	21	22	23	24
1st																								
2nd																								
3rd																								
4th																								
5th																								
6th																								
7th																								
8th																								
9th																								
10th																								
11th																								
12th																								
13th																								
14th																								
15th																								
16th																								
17th																								
18th																								
19th																								
20th																								
21st							Summer solstice																
22nd																								
23rd																								
24th																								
25th																								
26th																								
27th																								
28th																								
29th																								
30th																								

J

THE SEA

Average sea temperature in Celcius

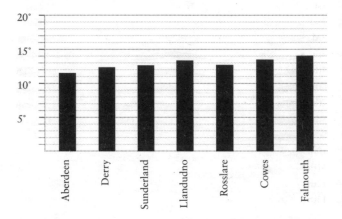

Spring and neap tides

Spring tides are the most extreme tides of the month, with the highest rises and the lowest falls, and they follow a couple of days after the full moon and new moon. These are the times to choose a low tide and go rock-pooling, mudlarking or coastal fossil-hunting. Neap tides are the least extreme, with the smallest movement, and they fall in between the spring tides.

Spring tides: 5th–6th and 19th–20th

Neap tides: 11th–12th and 27th–28th

Spring tides are shaded in black in the chart opposite.

June tide timetable for Dover

For guidance on how to convert this for your local area, see page 8.

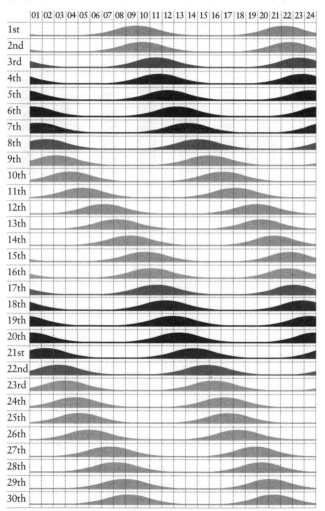

THE MOON

Moon phases

Full moon – 4th June, 04.42*

Last quarter – 10th June, 20.31

New moon – 18th June, 05.37

First quarter – 26th June, 08.49

*This full moon falls in the early hours of the morning. To catch it at its fullest during normal waking hours, view it the evening before.

Moonrise and set

Like the sun, the moon rises roughly in the east and sets roughly in the west. It also rises around 50 minutes later each day. Use the following guide to work out approximate moonrise times.

Full moon: Rises near sunset, opposite the sun, so in the east as the sun sets in the west.
Last quarter: Rises around midnight, and is at its highest point as the sun rises.
New moon: Rises at sunrise, in the same part of the sky as the sun (and so cannot be seen).
First quarter: Rises near noon, and is at its highest point as the sun sets.

Full moon

June's full moon is known as the Rose Moon or Dyad Moon.

New moon

This month's new moon, on the 18th, is in Gemini. Astrologers believe the new moon is a quiet, contemplative time before a phase of growth. Each new moon has its own energy, depending on the zodiacal sign that it is in. The Gemini new moon is said to rule communication.

Moon phases for June

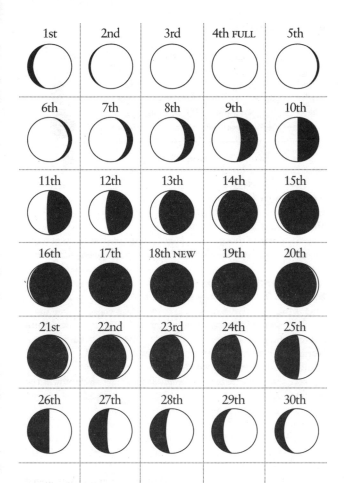

GARDENS

To enjoy this month

Ornamental: Roses, foxgloves, sweet peas, lady's mantle, peonies, lavender, elderflowers, bee orchid, lilies, ornamental alliums, Oriental poppies, *Knautia macedonica*, scabious, hardy geraniums, borage, astrantia, marigolds, clove pinks, pelargoniums, petunias, lobelia, fuchsia
Edible: Strawberries, gooseberries, blackcurrants, cherries, loganberries, raspberries, redcurrants, rhubarb, asparagus, globe artichokes, broad beans, peas and mangetout, new potatoes, chives, basil, mint

Gardening by the moon
The following is a guide to planting with the phases of the moon, according to traditional practices. It also works as a guide to the month's gardening for moon-gardening cynics, who can do these jobs whenever they wish during the month ahead.

First quarter to full moon: 27th May (from 16.22)–3rd and 26th (from 08.49)–3rd July (till 12.39)
This is the best time for sowing crops that develop above ground, but is bad for root crops. Plant out seedlings and young plants. Take cuttings and make grafts. Avoid any other pruning. Fertilise.
- Place brassica collars around the stems to keep out cabbage root fly, and net them to keep out cabbage white butterflies.
- Plant sweetcorn in a block, with plants about 35cm apart.
- Plant tomatoes into the ground or final pots in the greenhouse, with sturdy support. Nip out any sideshoots on cordon varieties. Feed weekly with a high-potash fertiliser.
- Direct sow beetroot, carrots, courgettes, cucumbers, French and runner beans, kale, peas, swedes, turnips and herbs.
- Sow Oriental leaves such as mizuna, mibuna, pak choi and mustard greens after the solstice.

Full moon to last quarter: 4th–10th (till 20.31)

A 'drawing down' energy. This phase is a good time for sowing and planting any crops that develop below ground: root crops, bulbs and perennials. Light is high but decreasing.

- Sow maincrop varieties of carrots. Sow autumn beetroot, swedes, turnips and spring onions.
- Pot up strawberry runners.
- Lift, divide and replant overcrowded clumps of spring bulbs.

Last quarter to new moon: 10th (from 20.31)–17th

A dormant period, with low sap and poor growth. Do not sow or plant. A good time though for pruning, while sap is slowed. Weeding now will check growth well. Harvest any crops for storage. Fertilise and mulch the soil. Garden maintenance.

- Stop cropping asparagus on the summer solstice, 21st June. Allow the plants to grow tall and ferny and water regularly with a balanced fertiliser.
- Net strawberries to protect them from birds.
- Earth up potatoes.
- Weed.
- Tie growing crops into supports.
- Inspect lilies for scarlet lily beetle and their larvae.
- Deadhead roses as they fade, to encourage more flowers.

New moon to first quarter: 18th–26th (till 08.49)

The waxing of the moon is associated with rising vitality and upward growth. Towards the end of this phase plant and sow anything that develops above ground. Prepare for growth.

- Pot on chillies and tomatoes into their final pots.
- Plant out bedding and hanging baskets. Water them every day and feed weekly.

Note: Where no specific time for the change between phases is mentioned, this is because it happens outside of sensible gardening hours. For exact changeover times, refer to the moon phase chart on page 133.

THE RECIPES

Bun of the month

Lemon, elderflower and poppy seed Sally Lunns

There are a couple of stories about the origin of the name Sally Lunn, a large and very light brioche-style bun originating in Bath, and still sold there in quantities. One legend is that the recipe was brought to Bath in 1680 by a young Huguenot refugee named Solange Luyon, recreating the festival breads she would have known in France. The locals, unable to pronounce her name, called her Sally Lunn, and the buns took on her name. There is no compelling evidence for Solange ever having lived, however. Another theory is that Sally Lunn is an anglicisation of *soleil et lune* – the French for sun and moon – after the bun's golden rounded top, and its white interior when split. It is in this spirit that it is our seasonal bun for the month of the summer solstice. This version has lemon and poppy seeds added to the dough and an elderflower glaze, some of the flavours of midsummer.

Makes 12
Ingredients
230ml milk
80g butter
40g clear honey
1 teaspoon lemon extract, or the zest of 2 lemons
420g strong white bread flour
1 teaspoon salt
1 sachet (7g) instant yeast
2 tablespoons poppy seeds
2 eggs
Elderflower cordial, to brush

Method

Grease a muffin tray. Gently heat the milk with the butter and honey in a saucepan until warm and melted. Remove from the heat and add the lemon extract or the zest. Allow to cool. In a large bowl, combine the flour, salt, yeast and poppy seeds. Make a well in the centre and crack in the eggs, mixing the egg and flour with a fork and then slowly adding the milk mixture. This is a very wet dough so may well need a little more flour to make it manageable, but add only as much as you need, as it should be a very light bun. Knead gently to get the measure of it, then a little more vigorously until you have a smooth and elastic dough.

Pop it back in the bowl, cover and leave in a warm place to grow plump and pillowy. After about 1 hour, knock the air out from the dough and shape into 12 small balls. Place each ball in a hole in the muffin tray. Cover and allow to rise again for 1 hour.

Meanwhile, preheat the oven to 180°C, Gas Mark 4 and, when the muffins have risen, bake them for about 15 minutes. Remove from the oven and brush them with elderflower cordial while they are still warm.

Chive *fritto misto* with lemon and anchovy mayonnaise

Chive flowers are edible and pretty. Break up the florets and strew them across salads, or encase them in a light, crisp batter and eat them hot. This recipe works beautifully with other sprigs of herb flowers and their leaves, too, such as sprigs of basil, oregano flowers and parsley leaves.

Serves 4 as a snack
Ingredients
For the mayonnaise:
4 anchovy fillets
Zest and juice of 1 lemon
3 tablespoons mayonnaise
For the *fritto misto*:
Sunflower oil, for frying
75g cornflour
75g plain flour
1 teaspoon salt
150ml sparkling water, chilled
20–30 chive stems

Method

Mash the anchovy fillets in a bowl with the lemon zest and juice. Mix in the mayonnaise and season to taste. Set aside.

Fill a saucepan a third full with oil and heat to 170°C – if you don't have a thermometer, drop a small piece of bread into the oil. It should fizz immediately and take 1 minute to brown. In a bowl, whisk together the cornflour, flour, salt and sparkling water. Dip the stems into it, and then, using tongs, carefully lower them into the hot oil, turning them until they are crisp and lightly browned. Drain on kitchen paper and serve with the mayonnaise.

THE ZODIAC

Gemini: 21st May–20th June

The sun begins the month in the same area of sky that holds the constellation of Gemini, the Twins, the 60th–90th degree of the zodiac. On the 21st of this month the sun will move into Cancer (see page 160).

> **Symbol:** The Twins
> **Planet:** Mercury
> **Element:** Air
> **Colour:** Yellow
> **Characteristics:** Chameleon-like, emotionally intelligent, optimistic, clever, communicative

Castor and Pollux were twins, of sorts. Castor was fathered by Tyndareus, their mother's husband, and Pollux by – you guessed it – Zeus (disguised as a swan this time). As adults, the twins were renowned for their horsemanship. In a raid to avenge the theft of their cattle, Castor was fatally struck by a spear. Pollux begged Zeus to share his own immortality with his brother, so Zeus put them in the sky as Gemini. The two brightest stars are called Castor and Pollux and the best time to spot Gemini is when it is in the opposite part of the sky from the sun six months from now, in December.

A FOLK SONG FOR GEMINI'S TWINS

'Two Young Brethren'
Traditional, arr. Richard Barnard

Here is a much simpler tale of brotherly exploits than that of Castor and Pollux, the twins of Gemini, in this month's zodiac constellation. This song is happily concerned with the annual cycle of agriculture, and two brothers' part in it, from ploughing and sowing the field, to bringing in the harvest, and then in autumn looking forward to doing it all again the next year.

Come all you jol-ly plough-men, come help me to sing, I'll sing in praise of you all, and if we don't la-bour, how shall we get bread? I'll sing to be mer-ry with - al.

Come all you jolly ploughmen, come help me to sing,
I'll sing in praise of you all,
And if we don't labour, how shall we get bread?
I'll sing to be merry withal.

Now here's two young brethren, two brothers and boys,
And two loving brothers so born.
For one was a shepherd, a tender of sheep,
The other a planter of corn.

Here's April, here's May and here's June and July
What pleasure to see the corn grow.
In August we will reap it, cut high and shear low,
And go down with scythes for to mow.

Then after we've laboured and reaped every sheaf
And gathered up every ear,
We've no more to do but to plough and to sow
To provide for the very next year.

NATURE

The pond in June

Water lilies are spreading themselves out across the surface of the water and may produce their first flowers this month. They may be the stars of the show but they are not the only flowers: water forget-me-not, water buttercups, water mint and marsh marigolds are all in flower. Solitary bees visit them for pollen but also make trips to the muddy edge of the pond for mud to seal their nests. The pond level may be dropping now if the weather is dry.

On still, warm days, the surface is busy with pond skaters, flicking and gliding, their feet splayed so that they do not break the surface tension, merely dimpling the water as they read it for vibrations that tell them prey is nearby.

The air above the pond is humming with life. Mayflies have been living as larvae in the mud at the bottom of the pond for up to three years, feeding on algae, strengthening themselves for this moment. The signal is a warm summer's day, whereupon they hatch, carry out their mating dance above the water, breed, lay eggs and then, as the sun sets, die, falling lifeless to the surface of the water. All of this in the space of a few hours, and from one of the earliest winged insects: they have been playing out this anticlimactic life cycle since before the dinosaurs.

Mosquito and midge larvae also hatch from the pond this month and many are snapped up by adult frogs – which can sit very still on lily pads or at the edge of the water and hook them in with their long, sticky tongues – or by spotted flycatchers or pipistrelle bats.

In larger ponds, this is often the month when ducklings will hatch and take to the water, fluffily following the leader along the duckweed paths. Their mother will teach them how to dive so as to forage at the bottom of the pond, which is now bristling with the larvae of craneflies, whirligig beetles, midges, gnats and mayflies – a nutritious scoop for a young duck. The other babies now finding their feet are the froglets that have completed their metamorphosis and are exploring the weedy and muddy edges of the pond.

Water lily

Water poppy

Water soldier

Frogbit

Water buttercup

July

 Wimbledon Women's Singles Final

 Wimbledon Men's Singles Final

Sea Sunday (Christian)

 Battle of the Boyne – bank holiday,
Northern Ireland

 St Swithin's Day (Christian/traditional)

 Al Hijra/Islamic New Year, start of the
Islamic year 1445, begins at sighting of the
crescent moon (Islamic)

23 Birthday of Haile Selassie (Rastafarian)

26 Tisha B'Av/Jewish day of mourning begins
at sundown (Jewish)

JULY AT A GLANCE

This is high summer. Evenings stretch, languidly. Trees wave their full and deep green crowns against blue skies. Music floats through open windows on warm breezes, reminding us that life really can be just this simple.

July holds an odd place in the ritual year. Almost every month has a host of major traditional celebrations attached to it which give hints as to what the atmosphere of the month would have been in the past, but in July there is almost nothing. Sometimes, however, absence can tell us plenty, and it is not that our farming ancestors were taking a well-earned breather at this time of year, kicking back and enjoying the sunshine. Quite the opposite. Traditional celebrations fell in the breathing spaces between the major jobs, and July was the time for one of the biggest of all: haymaking. This hugely labour-intensive job involved the whole community and was essential for preparing for winter, allowing animals to be kept alive when there was no fresh grass. It was brutally tough work that left no time to stop and put on a fête or a parade.

And so this is one moment in the year when the atmosphere of the month diverges from what it would have been in the past. You may well be working hard, but if you can possibly give yourself a little space to enjoy the luxury of the season, then take it, and relish that difference. Your ancestors would be proud to see how far you have come, sipping a glass of cold wine and laughing in the sun. Allow the ease of summer living to sink into your bones.

THE SKY AT NIGHT

Venus, which has been high in the sky through the spring evenings, now moves too close to the sun for us to see it well, and is finally lost in the sun's glare by mid-month. It will reappear as a morning star by the end of August. Meanwhile Jupiter becomes visible as a morning star from around mid-month, rising in the east about 01.30. It will become more prominent as the weeks go by.

7th: Close approach of the moon and Saturn. They rise together just before midnight in the east, then become lost in the dawn at about 04.30 in the south at an altitude of 27 degrees above the horizon.

12th: Close approach of the moon and Jupiter. They rise at around 01.20 in the east, then become lost in the dawn at about 04.30 in the east at an altitude of 31 degrees above the horizon.

28th–29th: Delta Aquarids meteor shower. An average shower which putters on from 12th July to 23rd August and produces up to 20 meteors per hour at its peak on the night of the 28th and early morning of the 29th. The nearly full moon will obscure all but the brightest trails.

J

THE SOLAR SYSTEM

The moon

Earth's moon is the largest satellite in the solar system relative to the size of its planet – its diameter is a little less than the distance across Australia from east to west. It is thought to have formed after the collision of a Mars-sized planet, Theia, with the earth, which sent debris flying out that was then caught in orbit and aggregated into the moon. This great cosmic piece of chance may well have helped create the circumstances that allowed life to evolve on earth. Early in its life, the moon was much closer in its orbit, and it is thought that its magnetic field acted as a shield, protecting the earth's delicate atmosphere from a particularly violent spell of solar winds. The moon's stronger gravitational pull-back then may have activated plate tectonics and seismic activity. These allowed the surface to cool, setting the earth apart from planets such as Venus, which has an uninhabitably hot surface. The gravitational pull of the moon also slowed the earth's rotation from eight-hour days to twenty-four-hour days, having a huge impact on the way life has evolved.

This month's zodiac sign, Cancer, is said to be ruled over by the moon. Cancer is considered the sign of homemaking and protection, which is neat considering the role the moon has played in protecting the earth from cosmic forces and helping to set the conditions that allowed life to evolve on our planet.

Sunrise and set

Haltwhistle, Northumberland

	01	02	03	04	05	06	07	08	09	10	11	12	13	14	15	16	17	18	19	20	21	22	23	24
1st																								
2nd																								
3rd																								
4th																								
5th																								
6th																								
7th																								
8th																								
9th																								
10th																								
11th																								
12th																								
13th																								
14th																								
15th																								
16th																								
17th																								
18th																								
19th																								
20th																								
21st																								
22nd																								
23rd																								
24th																								
25th																								
26th																								
27th																								
28th																								
29th																								
30th																								
31st																								

J

THE SEA

Average sea temperature in Celcius

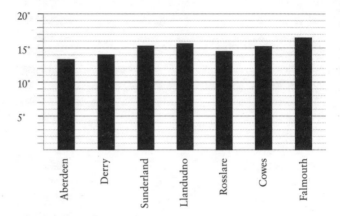

Spring and neap tides
Spring tides are the most extreme tides of the month, with the highest rises and the lowest falls, and they follow a couple of days after the full moon and new moon. These are the times to choose a low tide and go rock-pooling, mudlarking or coastal fossil-hunting. Neap tides are the least extreme, with the smallest movement, and they fall in between the spring tides.

Spring tides: 4th–5th and 18th–19th

Neap tides: 11th–12th and 26th–27th

Spring tides are shaded in black in the chart opposite.

July tide timetable for Dover

For guidance on how to convert this for your local area, see page 8.

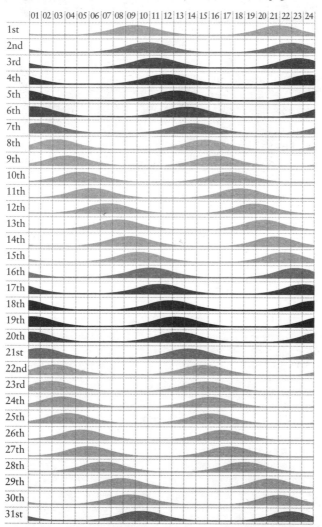

J

THE MOON

Moon phases

Full moon – 3rd July, 12.39

Last quarter – 10th July, 02.48

New moon – 17th July, 19.32

First quarter – 25th July, 23.07

Moonrise and set

Like the sun, the moon rises roughly in the east and sets roughly in the west. It also rises around 50 minutes later each day. Use the following guide to work out approximate moonrise times.

Full moon: Rises near sunset, opposite the sun, so is in the east as the sun sets in the west.
Last quarter: Rises around midnight, and is at its highest point as the sun rises.
New moon: Rises at sunrise, in the same part of the sky as the sun (and so cannot be seen).
First quarter: Rises near noon, and is at its highest point as the sun sets.

Full moon

July's full moon is known as the Wyrt (herb) Moon or the Mead Moon. It will be a supermoon, the first of four this year, which is noticeably bigger and brighter than other full moons.

New moon

This month's new moon, on the 17th, is in Cancer. Astrologers believe the new moon is a quiet, contemplative time before a phase of growth. Each new moon has its own energy, depending on the zodiacal sign that it is in, and the Cancer new moon is said to rule home, family and domesticity.

Moon phases for July

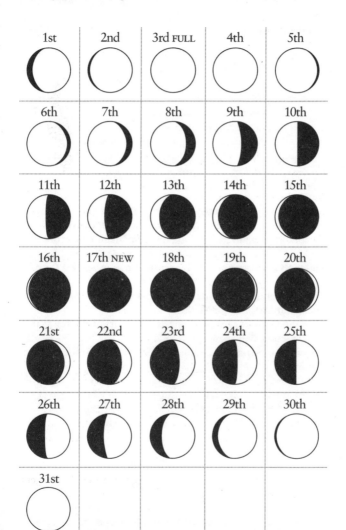

GARDENS

To enjoy this month

Ornamental: Sweet peas, lilies, day lilies, dahlias, marigolds, delphiniums, Oriental poppies, oxeye daisies, cosmos, sea holly, verbena, agapanthus, verbascum, honeysuckle, hydrangea, crocosmia, sweet rocket, love-in-a-mist, cornflowers, zinnias, larkspur, pelargoniums, petunias, lobelia, fuchsia
Edible: Blackcurrants, gooseberries, loganberries, raspberries, cherries, blueberries, French beans, runner beans, courgettes and courgette flowers, cucumbers, globe artichokes, peas, Florence fennel, new carrots, new potatoes, lettuces, mint, basil, dill, chives, marjoram, thyme, oregano

Gardening by the moon

The following is a guide to planting with the phases of the moon, according to traditional practices. It also works as a guide to the month's gardening for moon-gardening cynics, who can do these jobs whenever they wish during the month ahead.

First quarter to full moon: 26th June (from 08.49)–3rd (till 12.39) and 26th–1st August (till 19.32)

This is the best time for sowing crops that develop above ground, but is bad for root crops. Plant out seedlings and young plants. Take cuttings and make grafts. Avoid any other pruning. Fertilise.

- Sow French beans and peas for autumn harvesting.
- Plant out winter brassicas and protect with netting to keep off cabbage white butterflies.
- Sow salad leaves for autumn and winter harvesting: mustard greens, mizuna, mibuna, pak choi and chop suey greens, as well as kale, rocket, lettuces and Swiss chard.
- Take cuttings of woody herbs such as rosemary, sage and thyme.
- Feed everything.

Full moon to last quarter: 3rd (from 12.39)–9th

A 'drawing down' energy. This phase is a good time for sowing and planting any crops that develop below ground: root crops, bulbs and perennials. Light is high but decreasing.

- Sow beetroot for autumn harvesting. Continue to sow late maincrop carrot and turnip varieties and spring onions.
- Plant out leeks.
- Pot up strawberry runners for next year.

Last quarter to new moon: 10th–17th (till 19.32)

A dormant period, with low sap and poor growth. Do not sow or plant. A good time though for pruning, while sap is slowed. Weeding now will check growth well. Harvest any crops for storage. Fertilise and mulch the soil. Garden maintenance.

- Thin out the young fruits on apples and pears. Thin out grape bunches.
- Pinch out the sideshoots on tomatoes. Feed regularly with a high-potash feed, tying in stems as they grow.
- Pinch out the tops of climbing beans when they reach the tops of their frames. Spray the flowers of runner beans in hot weather to encourage pollination.
- Harvest onions, garlic and shallots for storage if you can.
- Weed regularly.
- Earth up potatoes.
- Prune cherries and plums if they need pruning.
- Cut the fruited canes of summer raspberries to the ground and tie in new growth.

New moon to first quarter: 17th (from 19.32)–25th

The waxing of the moon is associated with rising vitality and upward growth. Prepare for growth.

- You can sow, plant out or take cuttings of all of those things mentioned in the first quarter to full moon phase (see page 154). Ideally do this towards the end of this phase.

Note: Where no specific time for the change between phases is mentioned, this is because it happens outside of sensible gardening hours. For exact changeover times, refer to the moon phase chart on page 153.

THE RECIPES

Bun of the month

Cornish splits with thunder and lightning

Cornish splits are little sweet buns that are partly split –
usually on the diagonal rather than straight across – and filled
with whipped cream and jam. They are sometimes filled with
'thunder and lightning' (whipped cream and black treacle),
which evokes the sight of storm clouds building. This recipe
is offered here to mark St Swithin's Day on the 15th of this
month, when we hope for anything but. These days 'thunder
and lightning' is more often made with golden treacle,
presumably because of black treacle's uncompromising and
bitter flavour, but that feels like cheating, and doesn't create
the same effect at all. Instead, this version combines stewed
apples – apples are also associated with St Swithin – with the
black treacle, to soften it a little.

Makes 12
Ingredients
300ml milk
50g butter
25g caster sugar
400g strong white bread flour, plus extra for dusting
1 sachet (7g) instant yeast
1 teaspoon salt
1 teaspoon ground ginger
150ml double cream, to serve
For the treacle apples:
40g butter
40g black treacle
2 apples, peeled, cored and diced
1 teaspoon ground ginger

Method

Line a baking tray with parchment. In a saucepan, gently heat the milk, butter and caster sugar until melted, then allow to cool. In a large bowl, combine the flour with the yeast, salt and ground ginger, then add the cooled milk, forming a substantial dough. Turn out onto a floured surface and knead for 5 minutes until smooth and elastic. Pop back in the bowl, cover and leave in a warm place for about 1 hour until it has puffed up well. Knock back the dough and then form 12 small balls by rolling each firmly in your hands until it's relatively tidy. Place on the tray, cover again and leave for 30 minutes. Meanwhile, preheat the oven to 200°C, Gas Mark 6, and when the buns have risen, bake them for 10–15 minutes until lightly golden.

For the treacle apples, put the butter, black treacle, diced apples and ground ginger in a small saucepan, leaving it to bubble away for 5 minutes, stirring occasionally. Once the apple is soft, mash to a pulp and allow to cool.

Whip up the double cream. To serve, slice the buns in half on the diagonal and smooth over a dollop of cream, then top with the mashed treacle apples.

J

Meadowsweet champagne

Meadowsweet is in cloud-like flower now, wafting its scent of hay and honey from damp meadows and ditches. The flowers can be used in the same way as elderflowers, to make a fragrant, summery, sparkling and only very vaguely alcoholic drink.

Makes 5 × 1-litre bottles
Ingredients
20 heads of meadowsweet flowers
700g granulated sugar
2 tablespoons white wine vinegar
4.5 litres cold water
Zest and juice of 1 lemon

Method
Put all of the ingredients into a large bowl, saucepan or brewing bucket and stir well to dissolve some of the sugar. Loosely cover with muslin or a tea towel and leave to sit for 24 hours. Stir, and then strain into clean plastic screw-top bottles, filling each three-quarters full.

Unscrew every few days to prevent too much build-up of pressure. It will be ready to drink after 2 weeks. Serve chilled.

Apple, raspberry and basil fruit leather

Raspberries are at the height of their season now. If you have a glut of them, make some into fruit leather (puréed and dried fruit) for snacking and packed lunches. You can miss out the basil if you like, but it melds beautifully with all summer fruits, adding a herbal aniseed scent that only enhances their flavours.

Makes about 20 strips
Ingredients
300g apples, peeled, cored and chopped
650g raspberries
Juice of ½ lemon
Leaves of a small bunch of basil, finely chopped

Method

Preheat the oven to 140°C, Gas Mark 1. Line a baking tray with parchment. Put the apples, raspberries and lemon juice into a saucepan and cook for about 10 minutes, or until the fruit has broken down. Push the pulp through a sieve and return it to a clean saucepan. Bring it back to a simmer, to drive off as much moisture as you can without it sticking and burning. Remove from the heat and leave to cool for 5 minutes before stirring in the basil.

Spread the pulp across the baking tray, and put into the oven for at least 5 hours, until it is tacky to the touch. Switch off the heat and leave it in there for another 2 hours as the oven cools down. Use scissors to cut it into strips, then roll each up and store them in an airtight container.

THE ZODIAC

Cancer: 21st June–22nd July

The sun begins the month in the same area of sky that holds the constellation of Cancer, the Crab, the 90th–120th degree of the zodiac.On the 23rd of this month the sun will move into Leo (see page 183).

> **Symbol:** The Crab
> **Planet:** Moon
> **Element:** Water
> **Colour:** Silver-blue
> **Characteristics:** Emotional, intuitive, loyal, creative, spiritual, caring, resilient

Hera, Zeus' wife, was quite understandably almost permanently annoyed at Zeus' philandering with other women, and she took a strong and vengeful dislike to them and their children. She particularly hated Heracles, perhaps because he had been renamed specifically to try to mollify her – his name meant 'Hera's pride'. It didn't work. She spent a huge amount of time tormenting him in various awful ways. She drove him temporarily mad, leading him to kill his own children, and when he recovered he had to carry out twelve labours for his enemy Eurystheus (who was only his enemy because of earlier Hera troublemaking). One of these labours was to fight the Hydra, a fire-breathing monster with multiple heads that grew back as soon as they were chopped off. While Heracles was battling it, a crab latched onto his foot, and he had to turn his attention from the Hydra to crush the crab before continuing his fight. Hera rewarded the crab for its sacrifice by setting it in the heavens as the constellation Cancer. The best time to spot Cancer is when it is in the opposite part of the sky from the sun six months from now, in January.

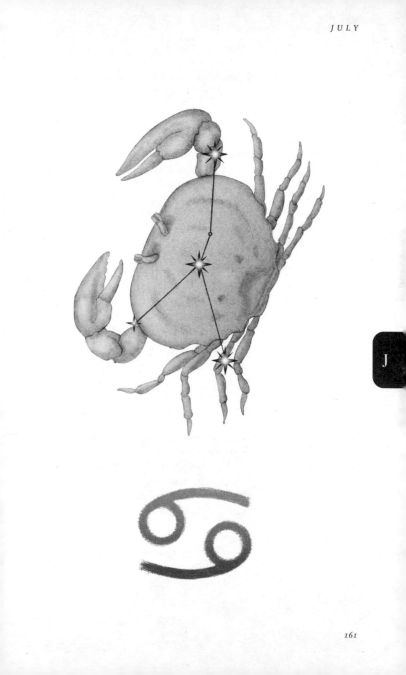

J

A FOLK SONG FOR CANCER'S CRAB

'The Crabfish'
Traditional, arr. Richard Barnard

The crab in this song is on its way to the cooking pot, but is stored in a different sort of pot in the meantime, with predictable results. Vulgar it may be (and there are far worse versions than this) but this is an extremely old song, originating as an Italian tale well known throughout Europe and the Middle East from at least 1400. It is here to represent the equally pesky crab of the story behind this month's zodiac sign, Cancer.

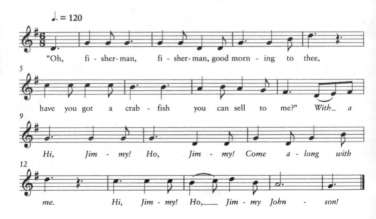

'Oh, fisherman, fisherman, good morning to thee,
Have you got a crabfish you can sell to me?'
With a Hi, Jimmy! Ho, Jimmy! Come along with me.
Hi, Jimmy! Ho, Jimmy Johnson!

'Oh, yes,' said the fisherman, 'indeed, sir, I do,
I have a little crabfish I can sell to you.'
With a Hi, Jimmy! Ho, Jimmy! Come along with me.
Hi, Jimmy! Ho, Jimmy Johnson!

I took the crabfish home and I thought he'd like a swim,
So I filled the chamber pot and put the fellow in.
With a Hi, Jimmy! Ho, Jimmy! Come along with me.
Hi, Jimmy! Ho, Jimmy Johnson!

My wife she got up in the middle of the night,
The crabfish and his claws, well, they gave her such a fright.
With a Hi, Jimmy! Ho, Jimmy! Come along with me.
Hi, Jimmy! Ho, Jimmy Johnson!

My wife grabbed a brush and I grabbed a broom,
We chased that poor old crabfish all around the room.
With a Hi, Jimmy! Ho, Jimmy! Come along with me.
Hi, Jimmy! Ho, Jimmy Johnson!

J

NATURE

The pond in July

In a dry summer the water level in the pond can drop dramatically by July, and the water becomes warm and thick with algae. But these changing conditions can be positive – many species lay their eggs in the exposed mud, for instance – and so there is no need to keep topping up the pond. Let it do its natural thing: rains will come soon enough. The warmth can make the mud at the bottom of the pond belch gas and the water can become stinky, but it will be rich with life.

Look out for dragonflies and damselflies mating. Some do so on the wing, flying linked together and with their long bodies bent towards each other to form a heart shape, while others will mate on a perch above the pond. Damselflies are slender and petite little lines of vibrant colour, whereas dragonflies are chunky and impossible to miss, buzzing noisily through the air and reaching the speed of up to 48 kilometres per hour. Damselflies rest with their wings against their bodies while dragonflies rest with them outspread. Both are brilliantly coloured – the jewel pins of the pond – and have astonishing manoeuvrability, as each wing is independently controlled, allowing them to hover, wheel and dart, snatching their prey out of the air. The females lay their eggs in the water or the damp mud around the pond edges, where they will hatch. They will live underwater as larvae or nymphs, feeding on tadpoles and water insects, moulting as they grow, for up to two years before they are ready to emerge and fly the pond themselves.

This month the first of the young newts, or 'efts', start to emerge from the pond, the rest following in August, and will begin their lives on land, hunting for slugs, worms and insects. It takes them up to four years to develop to sexual maturity; until then they will live in log piles and under rocks.

The adult toads that visited to breed will all have left the pond by the end of the month. There is no mass exodus – they leave singly at the end of the day, hiding among the long grasses and cloaking themselves in the night.

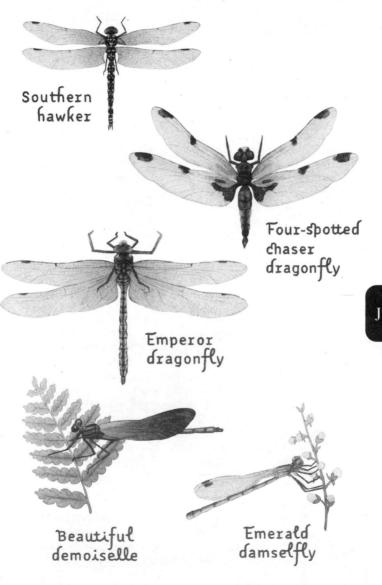

Southern
hawker

Four-spotted
chaser
dragonfly

Emperor
dragonfly

J

Beautiful
demoiselle

Emerald
damselfly

August

 Lammas (Christian) – blessing of the First Fruits harvest festival

 Lughnasadh (Gaelic/Pagan) – first harvest festival

 Summer bank holiday, Scotland. August bank holiday, Ireland

 The Assumption of Mary (Christian)

 FIFA Women's World Cup Final, Stadium Australia, Sydney

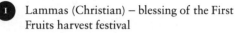 26th–28th: Notting Hill Carnival

Summer bank holiday, England, Wales and Northern Ireland

AUGUST AT A GLANCE

At the beginning of August falls Lammas, an English celebration with an Irish equivalent in Lughnasadh and a Welsh equivalent in Gwyl Awst (also known as Calan Awst). The English word comes from 'loaf mass', and the day marks the beginning of the first harvest, of grains and berries. The first sheaves of wheat are cut, and a loaf is made from the wheat and brought to church to be blessed. The months of harvest have begun.

The summer so far has been about growth, but now the energy has slowed and changed, to ripening and harvest. A few of the leaves on the lime trees turn lightly golden around their edges in treacherous hints of the autumn to come, and the papery, wheaten grass heads chirrup with grasshoppers. There are fewer flowers now, and the blackberry blossoms have turned to fat, shining fruits. It isn't yet time to think about winter and survival, but it is as if the year were just letting us know that it is there, on the horizon.

In fact, it is time now for gathering and feasting if you can, in these abundant, golden days, and in an echo of past Lammas celebrations. If you have a vegetable garden, it will now be overflowing with produce. Cook and bring people together to show off the fruits of your hard work. Or you might take this festival a little less literally, and take some time to think about all you have achieved during these past busy, light-filled months, and to pat yourself on the back for your hard work. No need to plan your next moves just yet – there will be plenty of time for introspection as the evenings draw in. But do take a moment to enjoy your achievements, to revel in them, to metaphorically nibble at a corner of the loaf, and be pleased with the beginnings of your harvest.

THE SKY AT NIGHT

Saturn reaches opposition by the end of the month, which is when it is at its closest to us, and so will be at its highest and brightest in our skies. Its rings will be 'open' and tipped towards us, so this will be the best time to view them with binoculars or a small telescope. There will not be a better opportunity to see the rings before 2026.

3rd: Close approach of the moon and Saturn. They rise at around 22.00 in the east. They reach their highest point at about 02.50 the next day in the south, at an altitude of 28 degrees. They then get lost in the dawn at about 05.00.

8th: Close approach of the moon and Jupiter. They rise just after midnight in the northeast. They become lost in the dawn at about 05.10 in the southeast at an altitude of 50 degrees above the horizon.

10th: Mercury at greatest eastern elongation. Mercury is the closest planet to the sun, and is hard to find if lost in the sun's glare. When it is at its furthest point from the sun in our sky, there is a chance to see it for a week or so. Spot it low in the western sky shortly after sunset.

12th–13th: Perseids meteor shower. One of the best meteor showers of the year, with up to 60 meteors an hour at its peak, is caused by the dust trail of comet Swift-Tuttle burning up as it hits our atmosphere. It runs from 17th July to 24th August, peaking in the night and early hours of the 12th and 13th. Best time for viewing is from about 21.00 to 03.00. The radiant (the point from which the meteor shower seems to emanate) will be at an altitude of about 40 degrees above the horizon in the northeast at midnight.

27th: Saturn at opposition, with its rings presenting to us. It will be at its highest at around midnight to 01.00. It will be high and bright in the week or so either side of this date.

30th: Close approach of the moon and a bright Saturn. They rise at around 20.10 in the southeast. They reach their highest point at 01.00 the following morning in the south at an altitude of 25 degrees above the horizon. They set in the southwest at around 05.30.

A

THE SOLAR SYSTEM

Saturn

Saturn's rings, which can be glimpsed this month via a good pair of binoculars or a telescope, are made up of billions of pieces of ice, each in its own orbit. Most of these are between 1 millimetre and 10 metres in size. They are constantly bashing against other pieces and fracturing, which is why they stay so reflective, and can be seen so well from earth. The ring system is around 400,000 kilometres wide, similar to the distance between the moon and earth. While the other gas giants – Jupiter, Uranus and Neptune – also have ring systems, Saturn's is by far the most spectacular. One theory is that it is the remnants of an icy moon that was broken apart in a collision, perhaps during the Late Heavy Bombardment (a possible event from several billion years ago, when multiple asteroids may have collided with early planets in our solar system).

When viewing Saturn from earth, the rings can be 'closed', with the rings edge-on so that they are hard to see; or they can be 'open', so that they form an oval that reflects the sun. Currently they are 'open'. With Saturn at opposition at the end of the month, this will be the best time to see them until 2026.

Sunrise and set
Haltwhistle, Northumberland

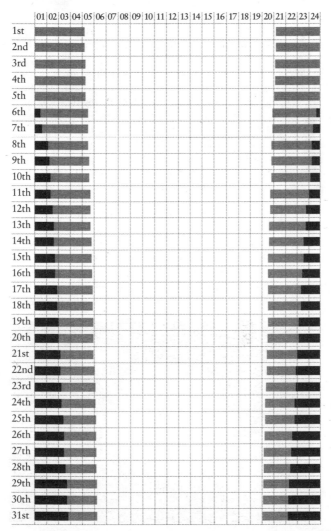

PLANETARY OPPOSITION

On 27th August, Saturn is at opposition, making the end of the month the perfect time to look for it in the sky. Opposition means that the planet is opposite the sun in our sky, and the term can only be applied to those planets that are further away from the sun than earth. Mercury and Venus, which are closer to the sun than earth is, cannot be at opposition. Setting them aside, picture all of the planets of the solar system orbiting around the sun, each at a different speed, with the earth also making one full orbit of the sun once every year. The further a planet is from the sun, the longer it takes to orbit the sun. Mars, next out from us, takes 687 days, or 1.9 years, to make a full orbit, Jupiter takes 11.9 years, Saturn 29.5 years, Uranus 84 years and Neptune 164.8 years. But all of this time, earth is whizzing around, once per year, in its relatively fast orbit. This means that there is a point once each year when we pass directly between each of these planets and the sun. It will change slightly each year, as those outer planets are moving, too.

Three things about this moment make them particularly brilliant for planet watching:

- We are as close as we are ever going to get to that planet, by dint of being in line with it, rather than around the other side of the sun.

- They are in our sky all night long. If we are sitting along the line that joins the sun and Saturn, then as any point on the earth moves through the day, the sun will be in our sky, as ever – but as the earth turns away from the sun into evening, Saturn will be opposite it, and so will rise in the east and by midnight will be high in our sky, always in the opposite part of the sky from the sun.

- The planet is fully illuminated. This is exactly the same phenomenon we see every month in the phases of the moon, albeit played out on a much larger scale: the moon is full when it is in the opposite part of our sky from the sun, and so this month we will be effectively viewing a 'full Saturn'. The only planet further out from earth that is not at opposition every year is Mars. This is because earth's

and Mars' orbits and speeds are quite closely matched,
so earth is not 'lapping' Mars as it does the other planets.
Mars is at opposition a little more than every two years – it
last occurred in December 2022 and will happen again in
January 2025.

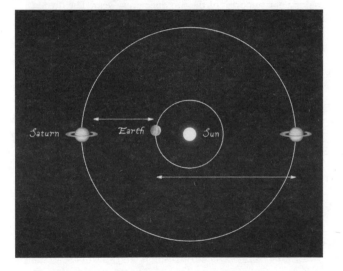

THE SEA

Average sea temperature in Celcius

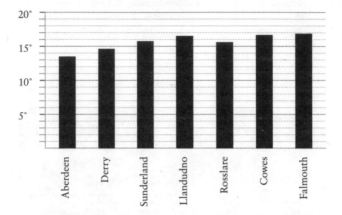

Spring and neap tides

Spring tides are the most extreme tides of the month, with the highest rises and the lowest falls, and they follow a couple of days after the full moon and new moon. These are the times to choose a low tide and go rock-pooling, mudlarking or coastal fossil-hunting. Neap tides are the least extreme, with the smallest movement, and they fall in between the spring tides.

Spring tides: 2nd–3rd and 17th–18th

Neap tides: 9th–10th and 25th–26th

Spring tides are shaded in black in the chart opposite.

August tide timetable for Dover

For guidance on how to convert this for your local area, see page 8.

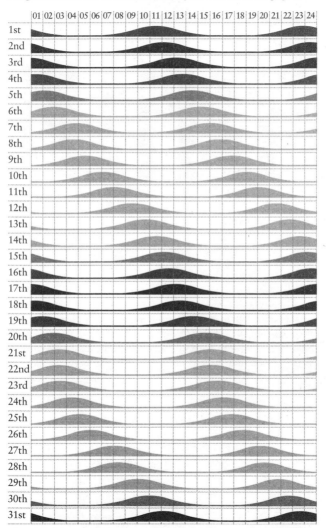

THE MOON

Moon phases

Full moon – 1st August, 19.32

Last quarter – 8th August, 11.28

New moon – 16th August, 10.38

First quarter – 24th August, 10.57

Full moon – 31st August, 02.36*

Moonrise and set

Like the sun, the moon rises roughly in the east and sets roughly in the west. It also rises around 50 minutes later each day. Use the following guide to work out approximate moonrise times.

Full moon: Rises near sunset, opposite the sun, so in the east as the sun sets in the west.
Last quarter: Rises around midnight, and is at its highest point as the sun rises.
New moon: Rises at sunrise, in the same part of the sky as the sun (and so cannot be seen).
First quarter: Rises near noon, and is at its highest point as the sun sets.

Full moon

The first full moon in August is known as the Grain Moon or the Lynx Moon. The second is the Wine Moon or Song Moon. When two appear in one month, the second is a 'blue moon'.

New moon

This month's new moon, on the 16th, is in Leo. The Leo new moon is said to rule confidence, energy, play and pleasure.

Moon phases for August

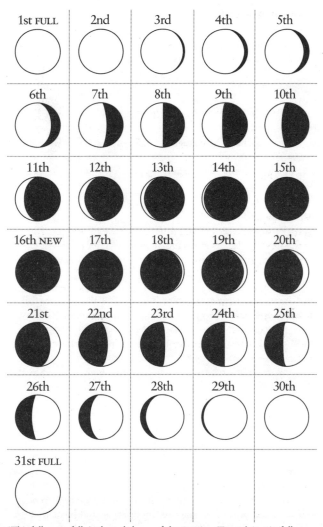

*This full moon falls in the early hours of the morning. To catch it at its fullest during normal waking hours, view it the evening before.

GARDENS

To enjoy this month

Ornamental: Dahlias, sunflowers, gladioli, crocosmia, cornflowers, crab apples, honesty seed heads, dill flowers, heleniums, sea holly, agapanthus, veronicastrum, verbena, clematis, marigolds, zinnias, nicotianas, phlox, lords-and-ladies, ornamental grasses
Edible: Plums, apples, pears, blackcurrants, elderberries, blueberries, loganberries, melons, raspberries, redcurrants, strawberries, cherries, tomatoes, sweetcorn, aubergines, French beans, runner beans, maincrop potatoes, Florence fennel, globe artichokes, sweet peppers, chilli peppers, marjoram, thyme, oregano, basil, mint

Gardening by the moon
The following is a guide to planting with the phases of the moon, according to traditional practices. It also works as a guide to the month's gardening for moon-gardening cynics, who can do these jobs whenever they wish during the month ahead.

First quarter to full moon: 26th July–1st (till 19.32) and 24th (from 10.57)–30th
This is the best time for sowing crops that develop above ground, but is bad for root crops. Plant out seedlings and young plants. Take cuttings and make grafts. Avoid any other pruning. Fertilise. Feed everything.
- Sow hardy annual flowers.
- Sow salad leaves for autumn and winter harvesting: mustard greens, mizuna, mibuna, pak choi and chop suey greens, as well as kale, rocket, lettuces and Swiss chard.
- Take cuttings of scented geraniums to overwinter.

Full moon to last quarter: 1st (from 19.32)–8th (till 11.28) and 31st–6th September
A 'drawing down' energy. This phase is a good time for sowing and planting any crops that develop below ground: root crops,

bulbs and perennials. Light is high but decreasing.
- Sow your last batch of carrots and turnips.
- Plant strawberry plants.
- Sow green manures to protect bare soil over winter: alfalfa, field beans, grazing rye or phacelia.

Last quarter to new moon: 8th (from 11.28)–16th (till 10.38)
A dormant period, with low sap and poor growth. Do not sow or plant. A good time though for pruning, while sap is slowed. Weeding now will check growth well. Harvest any crops for storage. Fertilise and mulch the soil. Garden maintenance.
- Deadhead dahlias to keep new flowers coming.
- Pinch out the sideshoots on cordon tomatoes and continue to feed regularly with a high-potash feed and to tie the stems in as they grow. Once a plant has produced four or five tresses of tomatoes (six or seven on some cherry types), 'stop' the plant by nipping out the tip, so encouraging ripening rather than growing.
- Pinch out the tops of climbing beans when they reach the tops of their frames, and spray the flowers of runner beans in hot weather to encourage pollination.
- Earth up potatoes. Weed regularly.
- Tie in the new growth of blackberries and boysenberries so their stems are horizontal or even slope downwards.

New moon to first quarter: 16th (from 10.38)–24th (till 10.57)
The waxing of the moon is associated with rising vitality and upward growth. Prepare for growth.
- You can sow, plant out or take cuttings of all of the things mentioned in the first quarter to full moon phase (see page 178). Ideally do this towards the end of this phase.

Note: Where no specific time for the change between phases is mentioned, this is because it happens outside of sensible gardening hours. For exact changeover times, refer to the moon phase chart on page 177.

THE RECIPES

Bun of the month

Blackberry and rose water Chelsea buns

Chelsea buns were first created as an Easter bun by bakers at the Chelsea Bun House in London in the early 18th century, but they became so popular that they are even now served in many bakeries across the country all year round. The sweet, enriched dough is rolled out flat and then a filling is scattered over it before it is rolled up and sliced like a Swiss roll, then laid on its side, baked and glazed. Here the filling is blackberry to mark the harvest of the first fruits at Lammas, which falls on 1st August.

Makes 12
Ingredients
250ml milk
90g butter
2 tablespoons brown sugar, plus extra for sprinkling
1 tablespoon rose water
400g strong white bread flour, plus extra for dusting
100g spelt flour
1 sachet (7g) instant yeast
1 teaspoon salt
1 egg
50g pistachios, roughly chopped

For the fruit filling:
50g blackberries
2 tablespoons caster sugar
3 tablespoons water
A splash of rose water

Method
Generously grease a 21cm round or square baking tin. In a
saucepan, gently heat the milk with 50g of the butter and
the brown sugar until dissolved and melted together. Remove
from the heat, add the rose water and allow to cool a little. In
a large bowl, combine the bread flour, spelt flour, yeast and
salt. Make a well in the centre and crack in the egg. Slowly
add the warm milk mixture and combine into a soft dough.
Turn onto a well-floured surface and knead, adding a smidgen
more flour if it's sticking. Bring together into a supple dough,
and pop the dough back in the bowl. Cover and place in a
warm place for about 1 hour, until the dough doubles in size.

Make the fruity filling by warming the blackberries
with the caster sugar, measured water and rose water in
a saucepan; mush it together. Sieve off the excess liquid,
reserving it for the glaze.

Knock the air out of the dough, turn it out onto a floured
surface and roll out into a 40 × 30cm rectangle. Melt the
remaining butter and brush over the dough. Spread the fruit
over the top and sprinkle with a handful of brown sugar and
the chopped pistachios.

Firmly roll up the long side to the top, slice into 12 pieces
and place in the baking tin – there should be just enough
room for the slices to spread out and touch sides. Cover and
leave to prove until puffed to the edges of the tin. Meanwhile,
preheat the oven to 200°C, Gas Mark 6, and when the buns
have puffed out, bake them for 25 minutes. Heat the reserved
liquid and brush this blackberry glaze over the buns while
they are still warm.

A

Wine and lemon braised artichoke

There's something hugely romantic about the thought of eating globe artichoke, though the prospect of preparing it may fill you with foreboding. It is well worth the effort to turn these bulbous beauties into this simple dish. It can then be used in a variety of ways, such as with ricotta to top pizza, combined with grains in a salad, or on fresh sourdough – or just eat it greedily in a sunny spot with oil.

Makes 16 pieces

Ingredients

Juice and zest of 1 lemon

2 garlic cloves, roughly chopped

3 tablespoons extra virgin olive oil

100ml white wine

2 sage leaves, finely sliced

2 medium globe artichokes

Method

Put the lemon juice and zest, garlic, olive oil, white wine and sage in a heavy-based frying pan over a low heat. Trim the artichoke stalks, removing the tough outer leaves until you reach the lighter, inner ones. (Depending on the size and age of your artichokes, you may need to use a small knife to trim back the outer leaves to get to the paler flesh.)

Cut the top 1cm off to see if there is an inner fuzzy 'choke' – if so, scoop it out with a spoon, making sure to get it all. Now cut each artichoke into 8 pieces and add to the frying pan. Cover with a lid and leave to steam, simmer and brown for about 20 minutes. Stir occasionally, adding a little water if it dries out but is not yet cooked through. The artichoke will be quite tangy when cooked, so taste it before adding any seasoning.

THE ZODIAC

Leo: 23rd July–22nd August

The sun begins the month in the same area of sky that holds the constellation of Leo, the Lion, the 120th–150th degree of the zodiac. On the 23rd of this month the sun will move into Virgo (see page 206). Mercury is also in retrograde from this day until 15th September.

> **Symbol:** The Lion
> **Planet:** Sun
> **Element:** Fire
> **Colour:** Orange, gold
> **Characteristics:** Courageous, warm, energetic, intense, high self-esteem, adventurous, optimistic

Leo commemorates another of the Labours of Heracles: the defeat of the Nemean Lion. The lion, whose golden fur repelled any mortal's weapons, had been sent by Hera to terrorise the people. Having failed to kill it with a bow and arrows, Heracles trapped the lion in a dark cave before stunning it with his club and strangling it with his bare hands. From then onwards, he wore its impenetrable pelt as armour. Zeus commemorated the feat by setting the lion into the sky as the constellation Leo. The best time to spot Leo is when it is in the opposite part of the sky from the sun six months from now, in February.

A

A FOLK SONG FOR LEO'S LION

'The British Lion'
Words traditional, music by Richard Barnard

This song, here for Leo's lion, is based on a Victorian 'broadside' from the Manchester Central Library broadside collection. Manchester was a great hub for the production of broadsides, which were cheap printed sheets of songs, sold on street corners in the 19th century to the urban poor. Written hastily, they often were about events in the news, and the best-loved songs might sell hundreds of thousands of copies.

Broadsides would mostly just contain words and no tune, as the tune would be widely known or borrowed. This is the case with 'The British Lion'. Composer Richard Barnard has created his own tune, in a Victorian song style.

The British Lion is a noble one,
And proud of his conscious might,
A terror to those he has made his foes
For he ever defends the right.
But yet so mild, a timid child
May approach him and never quail,
Pat him on his crown and stroke him down,
But beware how you tread on his tail.
Beware! Have a care! Beware! Have a care!
Beware how you tread on his tail!

Much is required to rouse his ire,
For he likes to lie down for a snooze.
No idle vaunt, no threat or taunt
Provokes his strength to use.
No bliss, he thinks, like forty winks,
Yet his vigilance will not fail,
For he sleeps with but one eyelid shut,
So beware how you tread on his tail.
Beware! Have a care! Beware! Have a care!
Beware how you tread on his tail!

Then up he bounds and his roar resounds
As he lashes his foaming sides.
His might he throws toward his foes
As he scatters them far and wide,
And great and small down, down they fall
Beneath his storm of hail,
And repeat to their cost, when all is lost,
That they trod on the Lion's tail.
Beware! Have a care! Beware! Have a care!
Beware how you tread on his tail!

A

NATURE

The pond in August

As the summer wears on, activity within the pond calms down from the fever pitch of spring and early summer. The water is warm and soupy, algae bloom and duckweed flourishes, and the growth around the pond is full and lush and deep green, but tinged with wear here and there. Grasshoppers stridulate from the long, golden grasses surrounding the pond, one of the final mating calls of the year. Swallows and swifts swoop over larger ponds, scooping their last drinks before setting off for Africa. There are more seed heads than there are flowers in the garden, and any ducklings and moorhen chicks have finally left their mothers and set out on their own. Everything is wrapping up now, breeding has taken place, babies have been raised. Minds are turning to winter.

But a few tadpoles may have been left behind. If you see any now, they may have 'Peter Pan syndrome' – that is, they have grown and grown but failed to develop in other ways. This is more likely to happen in a cold summer, and it is nothing to worry about. Tadpoles can overwinter, and if they make it through they will be well set up for next spring, able to grow to maturity ahead of the rest of the pack.

There is still plenty of life under the surface. The larvae and nymphs have mostly sunk down into the mud at the bottom of the pond, and there are fewer surface creatures darting about. Predatory larvae of bugs, dragonflies and beetles will still swim up to the surface to take drowning flies. Late summer rains will often arrive this month, starting to refill the pond at just the time it should be filled, refreshing both the pond and the garden. But before that, the rim of mud surrounding the pond makes it easy to see the footprints of the mammals that have been visiting it, such as voles, hedgehogs and foxes.

Marsh marigold

Yellow flag iris

Water forget-me-not

Marsh pennywort

Marsh cinquefoil

A

September

 Start of meteorological autumn

 Krishna Janmashtami – Krishna's birthday (Hindu)

 Enkutatash – Ethiopian New Year (Rastafarian)

 Rosh Hashana – Jewish New Year (Jewish) begins at sundown

 Ganesh Chaturthi – birth of Ganesh (Hindu)

 Autumn equinox, at 07.50 – start of astronomical autumn

 Mabon – harvest celebration (Pagan)

 Yom Kippur – Day of Atonement (Jewish) begins at sundown

 Prophet Muhammad's birthday (Islamic) begins at sighting of crescent moon

 Sukkot/First day of Tabernacles (Jewish) begins at sundown

 Michaelmas Day (Christian/traditional)

SEPTEMBER AT A GLANCE

The garden and countryside are full of fruits and seed heads suddenly, in the low late-summer light, the result of a summer of hard work growing and producing flowers, and of the bees' pollinating flights. The seeds will fall to the ground some time this month or next, and lie there on the ever-chilling soil, incubating, all of their potential contained until they can germinate and grow next spring.

Since midsummer we have known at the back of our minds that we were on the descent, but we could mostly ignore it. In September this awareness sharpens. There is a feeling now that the good living of summer must very soon come to an end, as we reach the autumn equinox and enter into the dark half of the year.

And so the energy of September centres around preserving – we collect up those seeds, label them and put them away, ready to be sown in the spring. And we bottle and make jam, and cook up chutneys with the harvests that didn't quite ripen. This month our seed boxes become packed with rustling and rattling paper bags, and our cupboards with glistening bottles and jars. We are behaving like the squirrels.

And we can choose to squirrel away less tangible things, too. Time will come during the winter to mull over thoughts and ideas in the darkness of our caves. The projects that have been forced to the back of our minds by the busy light months might now have the space to mature and incubate. And perhaps, like the seeds, they will be ready to burst forth when the light starts to return.

THE SKY AT NIGHT

Venus reappears in our sky this month, but it has now become a morning star, rising about an hour before the sun in the eastern sky. It will continue to rise earlier and earlier in the weeks ahead.

4th: Close approach of the moon and Jupiter. They rise at around 22.00 in the northeast and become lost in the dawn at about 06.00 the next day in the south at an altitude of 55 degrees above the horizon.

22nd: Mercury at greatest western elongation. Mercury is the closest planet to the sun, and this makes it hard to spot, as it is usually lost in the sun's glare. On the 22nd it will be the furthest it gets away from the sun in our sky, and so you may spot it for a week or so, low in the eastern sky a little before sunrise.

26th: Close approach of the moon and Saturn. They first appear in the dusk at about 19.20 in the southeast at an altitude of 10 degrees. They reach their highest point at 23.00 in the south at an altitude of 25 degrees. They set at about 03.30 in the southwest.

S

AUTUMN EQUINOX

Autumn equinox is associated with the harvest, and some harvest traditions take their timing from it. The Harvest Moon is the full moon nearest to the equinox, and Harvest Home or Ingathering, the day of end-of-harvest celebrations, was traditionally celebrated on the Sunday nearest the Harvest Moon (see also page 213).

This is because the autumn equinox is the moment we must start to say goodbye to summer, reap its rewards and prepare for winter. The earth is back at the point now where its tilt is side-on to the sun, just as it was at the vernal/spring equinox in March. This means that – momentarily – no hemisphere is favoured. Day and night are roughly even, all over the world, and the moment of the equinox occurs when the sun is directly overhead at the equator. But this time, the sun will slip a little further south, rather than north as it did in March. From there on, those of us in the northern hemisphere will have a slightly longer night than day, and this will continue until the winter solstice in December. For the southern hemisphere, this is, of course, the spring equinox, and the days will become longer and lighter now.

Common wisdom has it that on the equinox, day and night are exactly the same length, but this is not quite the case. It would be true if sunrise and sunset were measured from the moment that the centre of the sun rises over or sinks below the horizon, but we actually measure them from the moment the top edge appears or disappears. Thus, the equinox day is a little longer than its night, and its length varies slightly through the world. In autumn, the 'equilux', which is when day and night are exactly the same length, occurs a few days after the equinox.

Autumn equinox facts

Date: 23rd September
Time: 07.50
Altitude: At solar noon on 23rd September, the sun will reach 38 degrees in the London sky, 34 degrees in the Glasgow sky and 37 degrees in the Dublin sky.
Sunrise times: London 06.47, Glasgow 07.03, Dublin 07.12
Sunset times: London 18.57, Glasgow 19.14, Dublin 19.21
Sun's distance from earth: 150,128,000 kilometres

S

Mark the autumn equinox

- Make an autumn altar: a bowl of produce from the garden, a bunch of late-summer flowers in a jar, rosehips and acorns, orange candles.
- Plant tree seeds in pots of garden soil and keep them outside. Many tree seeds need a winter of cold before they will germinate. Look forward to the spring equinox, when you will hopefully see the first shoots.
- Fill a spray bottle with water, 1 teaspoon of vodka and 25–30 drops of autumnal essential oils, then shake it up and spritz it around. Try: cinnamon, cedarwood, cardamom, ginger.

Sunrise and set
Haltwhistle, Northumberland

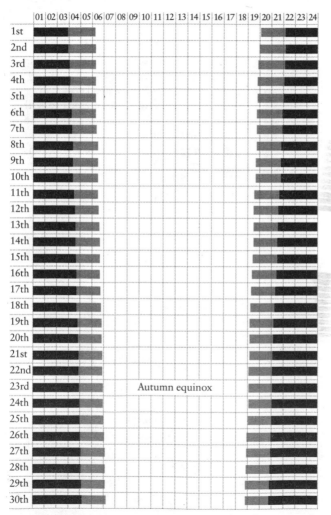

	01	02	03	04	05	06	07	08	09	10	11	12	13	14	15	16	17	18	19	20	21	22	23	24
1st
2nd
3rd
4th
5th
6th
7th
8th
9th
10th
11th
12th
13th
14th
15th
16th
17th
18th
19th
20th
21st
22nd
23rd								Autumn equinox
24th
25th
26th
27th
28th
29th
30th

S

THE SEA

Average sea temperature in Celcius

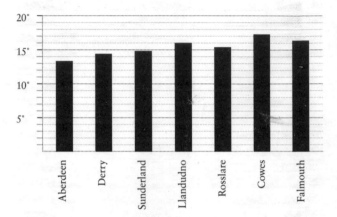

Spring and neap tides

Spring tides are the most extreme tides of the month, with the highest rises and the lowest falls, and they follow a couple of days after the full moon and new moon. These are the times to choose a low tide and go rock-pooling, mudlarking or coastal fossil-hunting. Neap tides are the least extreme, with the smallest movement, and they fall in between the spring tides.

Spring tides: 1st–2nd, 16th–17th and 30th–1st October

Neap tides: 7th–8th and 23rd–24th

Spring tides are shaded in black in the chart opposite.

September tide timetable for Dover

For guidance on how to convert this for your local area, see page 8.

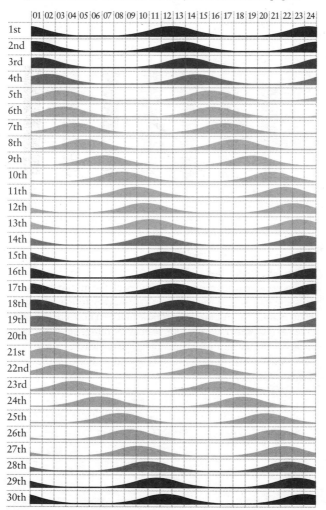

THE MOON

Moon phases

Last quarter – 6th September, 23.21

New moon – 15th September, 02.40

First quarter – 22nd September, 20.32

Full moon – 29th September, 10.58

Moonrise and set

Like the sun, the moon rises roughly in the east and sets roughly in the west. It also rises around 50 minutes later each day. Use the following guide to work out approximate moonrise times.

Full moon: Rises near sunset, opposite the sun, so in the east as the sun sets in the west.
Last quarter: Rises around midnight, and is at its highest point as the sun rises.
New moon: Rises at sunrise, in the same part of the sky as the sun (and so cannot be seen).
First quarter: Rises near noon, and is at its highest point as the sun sets.

Full moon

The full moon that falls closest to the autumn equinox each year is called the Harvest Moon, and this year it falls on 29th September. It will also be a supermoon, the final one this year.

New moon

This month's new moon, on the 15th, is in Virgo. Astrologers believe that the new moon is a quiet, contemplative time before a phase of growth. The Virgo new moon is said to rule routine and organisation.

Moon phases for September

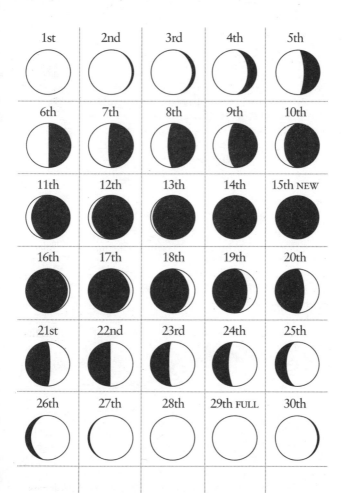

GARDENS

To enjoy this month

Ornamental: Dahlias, Michaelmas daisies, verbena, sunflowers, gladioli, spindle, scabious, echinacea, helenium, rudbeckia, phlox, cistus, autumn crocuses, crab apples, cyclamen, crocosmia, cannas, ornamental grass flower plumes
Edible: Blackberries, apples, pears, figs, damsons, loganberries, autumn raspberries, plums, redcurrants, tomatoes, aubergines, courgettes, chillies, peppers, sweetcorn, melons, pumpkins, winter squashes, maincrop potatoes, kale, thyme, mint, dill, oregano, marjoram, cobnuts, hazelnuts, walnuts, wild damsons

Gardening by the moon
The following is a guide to planting with the phases of the moon, according to traditional practices. It also works as a guide to the month's gardening for moon-gardening cynics, who can do these jobs whenever they wish during the month ahead.

Full moon to last quarter: 31st August–6th and 29th (from 10.58)–6th October (till 14.48)
A 'drawing down' energy. This phase is a good time for sowing and planting any crops that develop below ground: root crops, bulbs and perennials. Light is high but decreasing.
- Plant spring bulbs, daffodil bulbs in particular.
- Plant up forced hyacinth bulbs and paperwhites from late September to early October for Christmas flowers.
- Plant onion sets.
- Plant new perennials in your flower borders, and lift, divide and replant those that have finished flowering.

Last quarter to new moon: 7th–14th
A dormant period, with low sap and poor growth. Do not sow or plant. A good time though for pruning, while sap is slowed. Weeding now will check growth well. Harvest any crops for storage. Fertilise and mulch the soil. Garden maintenance.
- Lift maincrop potatoes and prepare them for storage.

- Pick apples and pears for storage.
- 'Stop' tomatoes (pinch out tips) once they have produced four or five tresses of tomatoes (six or seven on some cherry types). Continue to feed with a high-potash feed.
- Deadhead dahlias to prolong flowering.
- Order spring bulbs.
- Stop feeding houseplants and reduce watering towards the end of the month.
- Bring tender plants inside towards the end of the month.
- Save seed from favourite flowers or crops.

New moon to first quarter: 15th–22nd
The waxing of the moon is associated with rising vitality and upward growth. Plant and sow anything that develops above ground. Prepare for growth.
- You can sow, plant out or take cuttings of all of those things mentioned in the next moon phase, but towards the end of this phase is better.

First quarter to full moon: 23rd–29th (till 10.58)
This is the best time for sowing crops that develop above ground, but is bad for root crops. Plant out seedlings and young plants. Take cuttings and make grafts. Avoid any other pruning. Fertilise.
- Sow hardy annual flowers.
- Sow salad leaves and cover with cloches: Oriental greens, rocket, lettuces and Swiss chard.
- Sow sweet peas in the greenhouse or cold frame
- Plant winter bedding – foliage, pansies, violas and heathers – in pots and hanging baskets.
- Plant out wallflowers for a spring display.
- Take cuttings of scented geraniums to overwinter indoors.

Note: Where no specific time for the change between phases is mentioned, this is because it happens outside of sensible gardening hours. For exact changeover times, refer to the moon phase chart on page 199.

THE RECIPES

Bun of the month

Fig and tahini *challah* for Rosh Hashana

It is traditional for Jews to eat dense, egg-rich, braided *challah* for Shabbat each Saturday. At Rosh Hashana – Jewish New Year, which this year begins on the 15th of this month – sweet *challah* is made. It is always formed into a round, to represent the crown of God, the circle of life and the cycle of the year ahead. The sweetness often comes from currants mixed or folded into the dough, though here chopped figs are used instead. The bread is eaten dipped in honey, to symbolise the hope of a sweet year ahead.

Makes 1 large bun
Ingredients
1 sachet (7g) instant yeast
250ml warm water
100g strong white bread flour, plus extra for dusting
3 eggs
125ml olive oil
100g clear honey, plus extra to serve
1 teaspoon salt
400g bread flour
3 tablespoons light tahini
4 fresh figs, roughly chopped
3 teaspoons ground cardamom
5 tablespoons brown sugar
A splash of milk

Method

Line a 30cm-square baking tray with parchment. In a bowl, mix the instant yeast with the measured warm water and the strong white bread flour to make a smooth paste. Cover and leave somewhere warm for 1 hour until it gets bubbly. Uncover and crack in 2 eggs. Add the olive oil and honey, then mix well. Add the salt and bread flour, then form a rough dough.

Turn out onto a well-floured surface but try not to add too much more flour as the *challah* will not be as light. Knead for 5 minutes or until the ball has come together into a smooth dough. Pop the dough back in the bowl, cover and leave for an hour. Knock the dough back and, on a floured surface, roll it out into a rectangle about 1cm thick.

Drizzle the tahini over the dough and sprinkle over the figs, ground cardamom and brown sugar. Starting at one long edge, roll it up firmly into a giant sausage, then form that into a coil, tucking under the end. Transfer to the baking tray, cover and allow to prove for a further 45 minutes.

Meanwhile, preheat the oven to 200°C, Gas Mark 6. When the coil has risen, brush it with egg wash (1 egg beaten with a splash of milk), and bake for 30 minutes, until it has a good colour. Allow to cool before cutting and dipping in honey.

S

Romesco with grilled spring onions

In Catalonia, calçots – a type of green onion – are barbecued and served with Romesco sauce, a charred pepper sauce thickened with almonds. The season for this runs from December to April, but in the UK and Ireland peppers are gleaming and ripe in greenhouses now, and this is a lovely way to celebrate them. You will certainly be able to track down locally grown spring onions to eat along with them now. The sauce also goes well with grilled or barbecued seafood, or just with torn pitta to dip into the sauce. If you can, do the charring on a barbecue, as it tastes so much the better, but, if not, then all of the vegetables can be grilled.

Serves 4 as a starter
Ingredients
4 red peppers
2 garlic cloves
150g ground almonds
50g dry bread, cut into cubes
100g sun-dried tomatoes in oil
100ml extra virgin olive oil
20ml sherry vinegar
1 tablespoon smoked paprika
1 teaspoon cayenne pepper
Pinch of saffron, soaked in 1 tablespoon hot water
20–30 spring onions

Method

Put the peppers on a barbecue or under a high grill and blacken all over, turning them occasionally. Put them into a bowl and cover with clingfilm, leaving them to steam and cool for 10 minutes. Remove the flesh, discarding the blackened skins and seeds. Put the flesh, along with all of the other ingredients except for the spring onions, into a food processor. Pulse only briefly, ensuring the sauce keeps some texture and is not too smooth. Season to taste.

Put the spring onions on a barbecue or under the high grill, and cook until charred all over. Serve hot.

To eat, peel the blackened skins off the spring onions and scoop into the Romesco sauce.

S

THE ZODIAC

Virgo: 23rd August–22nd September

The sun begins the month in the same area of sky that holds the constellation of Virgo, the Virgin, the 150th–180th degree of the zodiac. On the 23rd of this month the sun will move into Libra (see page 228).

> **Symbol:** The Virgin
> **Planet:** Mercury
> **Element:** Earth
> **Colour:** Green
> **Characteristics:** Sophisticated, intelligent, kind, perfectionist, trustworthy, careful, organised

There are many different origin tales for the constellation of Virgo. In Roman mythology it is connected to Demeter, the Greek goddess of the wheat, with the sun in her constellation just as the wheat harvest begins. But the ancient Greeks associated it with the tale of two sisters, Parthenos and Molpadia, who had been left to watch over their father's wine. Unfortunately, while they were asleep the family's swine broke in and smashed the wine jar. Anticipating harsh punishment, the sisters tried to throw themselves off a cliff. Apollo, who had been the lover of their sister Rhoeo, saved them, and carried them to two different cities, where they were worshipped as goddesses. Parthenos died at a young age and Apollo set her in the stars as the constellation Virgo.

The best time to spot Virgo is when it is in the opposite part of the sky from the sun six months from now, in March.

Also in astrology: Mercury is in retrograde this month, from 23rd August to 15th September, which astrologers believe creates a period when communications break down, technology malfunctions, tempers fray and plans go awry (see page 82).

A FOLK SONG FOR VIRGO'S VIRGIN

'The Broomfield Wager'
Traditional, arr. Richard Barnard

There are a great many tales of 'maidenhood' and its loss throughout British and Irish folk songs, and many of them end in tragedy or, at the very least, an unfortunate marriage. This song, which is included here for the tragic figure of Parthenos, the woman behind the zodiacal sign of Virgo the Virgin, tells a happier tale than most. A rich young man lays a wager with a young woman that if she goes into the woods with him, she will return without her maidenhood. She takes the bet and wins it, still a virgin by the end of the day, and richer too.

Versions of this song go back around 700 years and folk song collectors found it throughout the English countryside, most often in the southwest of England, but also in Scotland and North America.

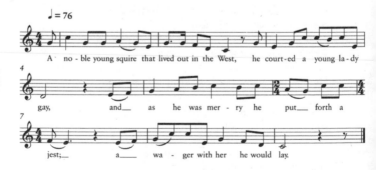

A noble young squire that lived out in the West,
He courted a young lady gay,
And as he was merry he put forth a jest;
A wager with her he would lay.

She asked him the wager, and this he did reply,
'I'll give you five hundred pounds and ten
If a maid you will go to the bonnie blooming bush
And a maid you return home again.'

Next morning she went to the bonnie blooming bush
Determined his money for to keep
And when she got down to the bonnie blooming bush
There she found her true love fast asleep.

Three times she did walk round the place where he lay,
Three times there she walked all around.
Three times did she kiss of his red and rosy lips
As he lay there asleep upon the ground.

Upon his right hand a gold ring she then secured
A ring from her own fingers fair,
That when he awoke he would be at once assured
That his lady and love had been there.

He wakened and saw the gold ring upon his hand
And back across the fields he did run.
She said, 'I have been to the bonnie blooming bush
And a maid I have returned back again.

Be cheerful, be cheerful and pray, do not complain,
For now it is as clear as the sun
The money you promised, the money now is mine;
Your wager I fairly have won.'

S

NATURE

The pond in September

The last days of summer are here and on the pond it may seem as if not that much has changed. Foliage is still lush. Brooklime, water forget-me-not and duckweed almost cover the pond, and there are even a few water lilies blooming. But the energy has changed and it is quiet on the pond, with breeding activity over.

Dragonflies will not fly for much longer. Having completed their life cycle, the adults will die when the weather turns colder. Most of the eggs that they have laid on the water have sunk into the mud at the bottom of the pond and now enter diapause, a state of suspension in which they will stop developing. Some will hatch and overwinter as larvae, but food is scarce in winter and they are vulnerable to starvation or predation.

While toads make their migrations to the pond in spring, newts do it this month. These are the newts that have just reached sexual maturity. They have been living on land under rocks and in log piles and have been eating slugs, worms and insects since crawling from the pond up to seven years previously. Incredibly sensitive and picky, they are able to sniff out and avoid ponds containing fish, while choosing those with the perfect acidity levels to give their offspring the best chance of success. They will overwinter in the pond, ready to breed next spring and summer.

Around the pond the first twinges of gold are coming into the leaves. Meadow brown butterflies settle and sun themselves but they, too, will not be around for much longer. The first daddy-longlegs are flitting around in the late summer sun. Wasps have been expelled from their nests by their queens as they prepare to overwinter, and they will soon die when the weather chills, but for now they search for sweetness in the garden's fallen fruits. The grass is long, the wildflowers are seeding, and birds, hedgehogs, mice and voles gorge on them and drink from the edges of the pond, their footprints left in the mud at the edges. In their various ways, every creature is now preparing for the colder months ahead.

Smooth newt

Palmate newt

Great crested newt

October

1   Harvest Home/Ingathering (traditional)

1   Start of Black History Month

1   Start of English pudding season

11   Old Michaelmas Day (Christian) – the day the Devil fell from heaven, landed on a bramble and cursed its fruits, so blackberries are traditionally not eaten after this date

21   Apple Day

29   British Summer Time and Irish Standard Time end. Clocks go back one hour at 02.00

30   October bank holiday, Ireland

31   Hallowe'en

OCTOBER AT A GLANCE

The mists really do start to roll in with the beginning of this beautiful month, the result of the ground chilling overnight as nights lengthen and the northern hemisphere tips further away from the sun. The leaves are turning brilliant shades of red and gold, and everything is ripening and closing down for the cold months ahead.

This is the month that we really start to feel the coming winter, nudged along rather rudely by the hour change on the 29th. We can be clinging to the last vestiges of hope that winter won't be too bad, that the nights aren't perhaps as long and cold as we have remembered, and then…bump. The hour changes and our day is cut short, our commute plunged into darkness.

It is hard not to resent this in the moment that it happens, and in the following days as we wrestle with this new dose of darkness just when we need it least. We always want the last dregs of summer to carry on. But try not to fight it. The dark and cold are here for a good reason. Look around you and you will see that everything in the natural world is closing down – you can take the darkness as a cue to do a little of this yourself. We need rest times as well as active times in order to stop ourselves from burning out, and this moment is as strong a signal to rest as you are ever going to receive, just as it is to the plants, the mammals and the insects. They know what to do, but we need to teach ourselves. A suggestion: treat yourself to some new pjs if possible, get into them as soon as you get home in the evening, perhaps after a hot scented bath, and then snuggle on the sofa with candles and crumble, and feel your shoulders dropping. Repeat.

THE SKY AT NIGHT

Venus reaches its highest altitude as a morning star mid-month, rising at about 03.00, some four hours before the sun. Orion, the Hunter, and Sirius, the Dog Star, will be high in the sky by mid-month, from 03.00. Towards the end of the month there will be a partial lunar eclipse to look out for.

1st: Close approach of the moon and Jupiter. They rise at 20.00 in the northeast and reach an altitude of 55 degrees above the horizon at 03.00 the next day in the south.

10th: Close approach of the moon and Venus. They rise at around 03.30 in the east.

14th: Annular solar eclipse, not visible from UK and Ireland.

21st–22nd: Orionids meteor shower. Runs from 2nd October to 7th November, peaking on the night of the 21st and early morning of the 22nd, when it will produce up to 20 meteors per hour. It occurs when the earth moves through the dust trail left by Halley's Comet, which burns up as it hits our atmosphere.

23rd: Mercury at greatest western elongation. It's hard to spot, but around the 23rd it will be the furthest from the sun so you may see it low in the eastern sky before sunrise.

24th: Close approach of the moon and Saturn. They first appear in the dusk at around 18.10 in the southeast at an altitude of 13 degrees above the horizon.

28th: Close approach of the moon and Jupiter. They rise at around 18.20 in the northeast. They reach their greatest altitude of 53 degrees above the horizon at about 00.50 in the south.

28th: Partial lunar eclipse, visible throughout Europe, Asia, Africa and Western Australia. The moon will enter earth's partial shadow, the penumbra, at 19.00 and will enter the full shadow, the umbra, by 20.30. A part of the moon will increasingly darken until reaching the maximum extent at 21.15. The moon will leave the umbra by 21.50 and the penumbra by 11.30. Only a small area will be darkened.

O

THE SOLAR SYSTEM

Neptune

Neptune is the furthest-known planet from the sun, up to 4,540,000,000 kilometres away from it. The sun's light on Neptune is 900 times weaker than it is on earth, resulting in a permanent dim twilight. In theory the planet could be seen with an amateur telescope if you used software to find out exactly where to look and had an extremely dark sky, but in reality it is so dim that this is near impossible. Bear in mind that it is the only planet that was not discovered by the naked eye or telescope, but was instead predicted by apparent anomalies within Uranus' orbit. However, this month it is just past opposition (see page 172), so it is at least at its closest and brightest now.

Named after the Roman god of the sea, Neptune is deep blue in colour and has a rocky core of nickel and iron, a sea of water, ammonia and methane ice, and a layer of hydrogen, helium and methane clouds. These gassy clouds race around the planet at up to 2,400 kilometres per hour, nearly twice the speed of sound – the solar system's fastest winds.

Sunrise and set

Haltwhistle, Northumberland

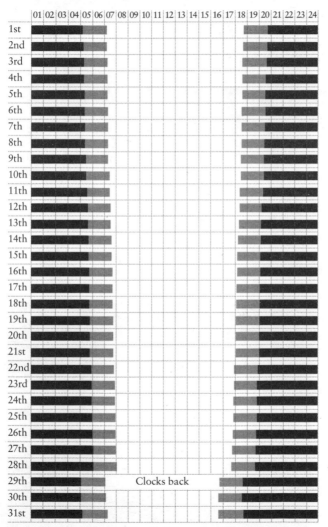

Clocks back

British Summer Time and Irish Standard Time end on 29th October at 02.00
and this has been accounted for above.

THE SEA

Average sea temperature in Celcius

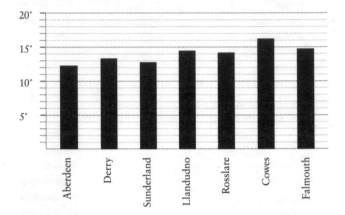

Spring and neap tides

Spring tides are the most extreme tides of the month, with the highest rises and the lowest falls, and they follow a couple of days after the full moon and new moon. These are the times to choose a low tide and go rock-pooling, mudlarking or coastal fossil-hunting. Neap tides are the least extreme, with the smallest movement, and they fall in between the spring tides.

Spring tides: 30th September–1st, 15th–16th and 29th–30th

Neap tides: 8th–9th and 23rd–24th

Spring tides are shaded in black in the chart opposite.

October tide timetable for Dover

For guidance on how to convert this for your local area, see page 8.

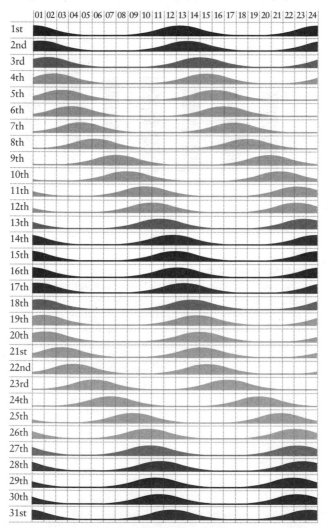

THE MOON

Moon phases

Last quarter – 6th October, 14.48

New moon – 14th October, 18.55

First quarter – 22nd October, 04.29

Full moon – 28th October, 21.24

Moonrise and set

Like the sun, the moon rises roughly in the east and sets roughly in the west. It also rises around 50 minutes later each day. Use the following guide to work out approximate moonrise times.

Full moon: Rises near sunset, opposite the sun, so in the east as the sun sets in the west.
Last quarter: Rises around midnight, and is at its highest point as the sun rises.
New moon: Rises at sunrise, in the same part of the sky as the sun (and so cannot be seen).
First quarter: Rises near noon, and is at its highest point as the sun sets.

Full moon

The full moon in October is known as the Hunter's Moon or Blood Moon.

New moon

This month's new moon, on the 14th, is in Libra. Astrologers believe that the new moon is a quiet, contemplative time before a phase of growth. Each new moon has its own energy, depending on the zodiacal sign that it is in, and the Libra new moon is said to rule harmony and diplomacy.

Moon phases for October

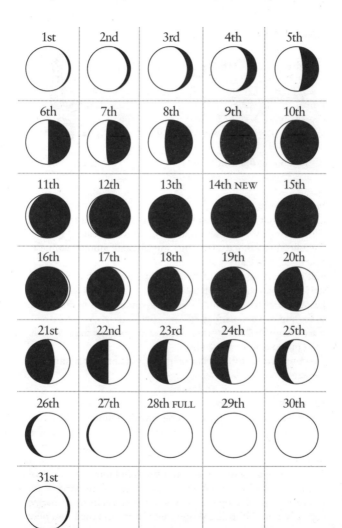

GARDENS

To enjoy this month

Ornamental: Michaelmas daisies, dahlias, chrysanthemums, rudbeckia, helenium, autumn leaves, seed heads, crab apples, rosehips, haws, rowan berries, ornamental grasses
Edible: Medlars, quinces, apples, pears, ceps, chanterelles, sloes, tomatoes, aubergines, chillies, peppers, Brussels sprouts, squashes, pumpkins, chervil, parsley, coriander, bay, rosemary,

Gardening by the moon
The following is a guide to planting with the phases of the moon, according to traditional practices. It also works as a guide to the month's gardening for moon-gardening cynics, who can do these jobs whenever they wish during the month ahead.

Full moon to last quarter: 29th September (from 10.58)–6th (till 14.48) and 29th–4th November
The waning moon is traditionally associated with a 'drawing down' energy, and this phase is a good time for sowing and planting any crops that develop below ground: root crops, bulbs and perennials. Light is high but decreasing.
- Plant up hyacinth, paperwhite and hippeastrum bulbs in early October for Christmas flowers.
- Plant bulbs: daffodils, crocuses, scilla, fritillaries and irises.
- Plant lilies and ornamental alliums.
- Plant garlic cloves and overwintering onion sets.
- Plant rhubarb crowns and bare-root fruit bushes.
- Plant new grapevines, peaches and nectarines.
- Plant new herbaceous perennials, and lift, divide and replant those that have finished flowering.

Last quarter to new moon: 6th (from 14.48)–14th
A dormant period, with low sap and poor growth. Do not sow or plant. A good time though for pruning, while sap is slowed. Weeding now will check growth well. Harvest any crops for storage. Fertilise and mulch the soil. Garden maintenance.

- Lift maincrop potatoes and prepare them for storage.
- Lift and store beetroot and turnips. Leave carrots and parsnips until needed.
- Cut pumpkins and winter squashes on a fine day and leave to 'cure' in the sun, then store them somewhere frost-free.
- Pick apples and pears for storage.
- Cover salad leaves and leafy vegetables with cloches to keep them in good condition through winter.
- Dismantle supports for runner beans, peas and tomatoes.
- Pick green tomatoes for chutney or put them into paper bags in a dark place where they can ripen slowly.
- Earth up Brussels sprouts, other brassicas and leeks.

New moon to first quarter: 15th–21st
The waxing of the moon is associated with rising vitality and upward growth. Plant and sow anything that develops above ground. Prepare for growth.
- You can sow, plant out or take cuttings of all of those things mentioned in the next moon phase, but towards the end of this phase is better.

First quarter to full moon: 22nd–28th
This is the best time for sowing crops that develop above ground, but is bad for root crops. Plant out seedlings and young plants. Take cuttings and make grafts. Avoid any other pruning. Fertilise.
- Sow broad beans, direct or in pots in the greenhouse.
- Sow sweet peas and hardy varieties of pea in pots in a greenhouse or cold frame.
- Plant out spring-flowering wallflowers, violas and forget-me-nots.

Note: Where no specific time for the change between phases is mentioned, this is because it happens outside of sensible gardening hours. For exact changeover times, refer to the moon phase chart on page 221.

THE RECIPES

Bun of the month

Barmbrack buns

Barmbrack is a sweet tea bread associated with Hallowe'en in Ireland, where various objects are baked into it for fortune-telling, similar to Epiphany *fève* traditions (see page 22), with the fate of the finder suggested by the object. A dried pea means you will not marry this year, a matchstick signifies you will have a year full of disputes, a cloth means you will be poor, a coin indicates you will be rich and a ring signifies you will marry within the year. Barmbrack is traditionally a large round loaf and is now made with baking soda, as most Irish traditional breads are, but it is possible that 'barm' comes from the froth of fermented ale, which was once used as yeast to make bread dough rise. This version uses a yeasted dough combined with barmbrack fruits and flavours, which allows you to make smaller buns, perhaps each containing a charm.

Makes 12
Ingredients
1 cup of strong black tea
100g currants
140g sultanas
80g mixed peel
230ml milk
80g butter
50g dark brown sugar
300g spelt flour
100g strong white bread flour, plus extra for dusting

| 1 teaspoon salt |
| 1 sachet (7g) instant yeast |
| 2 teaspoons ground nutmeg |
| 2 teaspoons ground cinnamon |
| Salted butter, to serve |

Method

Grease a muffin tin and put the strong cup of black tea into a mixing bowl. Weigh out the currants, sultanas and mixed peel, and soak in the tea for about 20 minutes. In a saucepan, gently heat the milk, butter and dark brown sugar until melted together; allow to cool. Combine the spelt flour, bread flour, salt, yeast, nutmeg and cinnamon in a large bowl. Add the cooled milk and mix to form a dough. Drain the fruits that have been soaking, discarding the tea, then add them to the dough and knead to combine.

Turn out onto a floured surface and knead for 5–10 minutes. Pop the dough back in the bowl, cover and leave in a warm place for at least 1 hour.

Meanwhile, preheat the oven to 180°C, Gas Mark 4. When the dough has risen, form 12 smooth little balls and pop one in each hole in the muffin tin. Bake for 15 minutes, then allow to cool. Serve with lashings of salty butter.

O

Hazelnut pakoras with smoky carrot dip

Peanut pakoras are a South Indian snack particularly popular during monsoon season. As we are entering our own rainy season this month, here is a version using crunchy hazelnuts, which are maturing in the hedgerows now.

Makes about 12

Ingredients

For the smoky carrot dip:

3 large carrots, cut in rounds

1 large red onion, coarsely chopped

2 garlic cloves

1 tablespoon smoked paprika

½ teaspoon ground cinnamon

1 teaspoon salt

2 tablespoons olive oil, plus extra to serve

2 tablespoons light tahini

Handful of dill, chopped

For the pakoras:

75g hazelnuts

100g buckwheat flour

1 tablespoon salt

½ teaspoon cayenne pepper

1 teaspoon cumin seeds

Handful of fresh coriander, finely chopped

Juice of 1 lemon

150ml water

200ml rapeseed oil

Method

For the dip, preheat the oven to 220°C, Gas Mark 7. Put all the dip ingredients, apart from the tahini and dill, in a roasting dish and roast for about 20 minutes. Remove from the oven and allow to cool a little before tipping into a bowl, combining with the tahini and dill and blitzing with a stick blender until smooth. Transfer to a serving bowl, top with a little more oil and set aside.

For the pakoras, toast the hazelnuts in a hot frying pan for a couple of minutes, shuffling them about to avoid burning. Allow the nuts to cool and then chop coarsely. Place the flour, salt, cayenne, cumin seeds and coriander in a large bowl, and mix. Using a whisk, add the lemon juice and measured water bit by bit. Once it is smooth, stir in the toasted hazelnuts.

Heat the rapeseed oil in a small saucepan over a medium heat. Test whether the oil is hot by adding a tiny dot of the batter, it should sizzle immediately. When the oil is bubbling away, add small spoonfuls of the mixture, turning them in the oil so they cook evenly. This should take around 2–3 minutes per batch. Remove and drain on kitchen paper, and cook the rest in batches. Serve warm with your smoky carrot dip.

O

THE ZODIAC

Libra: 23rd September–22nd October

The sun begins the month in the same area of sky that holds the constellation of Libra, the Scales, the 180th–210th degree of the zodiac. On the 23rd of this month the sun will move into Scorpio (see page 250).

> **Symbol:** The Scales
> **Planet:** Venus
> **Element:** Air
> **Colour:** Blue
> **Characteristics:** Diplomatic, loyal, well-balanced, caring, imaginative, artistic, charming, beautiful

Libra is the only constellation in the zodiac that represents an object rather than a person or creature. Originally the stars within it were not regarded as a constellation of their own but were the pincers on Scorpio. However, the Romans, possibly around the second century CE, decided it was a constellation and warranted its own name. It is known as the Scales of Justice, which were held by Themis, a Titaness and Zeus' second wife (and sister, but we won't dwell). She personifies justice, law, fairness and divine order. The image of Lady Justice, blindfolded and holding the scales, comes from Themis. The best time to spot Libra is when it is in the opposite part of the sky from the sun six months from now, in April.

O

A FOLK SONG FOR LIBRA'S SCALES

'Georgie'
Traditional, arr. Richard Barnard

Libra's scales represent fairness and the workings of justice, but historically the two have not always gone hand in hand. This angry and beautiful folk song takes us through the court system as a man is condemned to be hanged for the killing of the king's deer. It is told through the tale of his sweetheart, who is begging the judge for mercy, to no avail.

George Stoole, a Northumbrian robber who was executed in 1610, is one possible contender for the inspiration for Georgie. However, it seems as likely that the song is even older than that, having possibly emerged at the end of the medieval period, when the forests were full of outlaws on the run from the oppressive feudal system.

As I walked o-ver Lon-don Bridge one mis-ty morn-ing ear - ly, O, there I saw_ a fair young maid la - ment - ing for_ her Geor - gie.

As I walked over London Bridge
One misty morning early,
O, there I saw a fair young maid
Lamenting for her Georgie.

'Go saddle me my best black horse,
And saddle it quite swiftly,
That I may ride to the castle fair
And beg for the life of Georgie.

He never stole any horse nor cow
Nor done any harm to any
But he stole sixteen of the king's white steeds
and sold them in Bohenny.'

The judge looked over his left shoulder
As he spoke unto the lady.
He said to her 'You've come too late,
The man's condemned already.

He's walked all down the long lined streets,
He's bade farewell to many;
He's bade farewell to his own true love
Which grieved him more than any.'

'I wish I was on yonder hill
Where he was caught so cruelly,
With sword and pistol by my side
I'd fight for the life of Georgie.'

O

NATURE

The pond in October

Falling leaves are drifting onto the surface of the pond now, and the growth around the pond is dying back, taking on shades of gold, brown and red. The reeds and rushes are bowed from rain and wind, and spiders' webs are strung between them. The pond is filling up again with autumn rains and could be brimful by the end of the month.

This month frogs will go into their version of hibernation, called brumation, a period of dormancy in which they will shut down their bodies to preserve energy. A frog absorbs oxygen in three different ways: through the lungs, as you might expect; through the lining of the mouth; and via the skin, which must be moist at all times. Some will actually hibernate at the bottom of the pond, digging into the mud and sitting out the coldest weather. Others will create their hibernaculas (winter quarters) in the mud at the edge of the pond. Still others, including the year's juveniles, will find nooks and crannies in log or rock piles nearby – anywhere they can stay moist and keep off the worst of the cold until spring. You might spot a particularly fat frog at this time of year. This will be a female, already full of spawn, which she will carry all through the winter, ready for breeding time next February or March. Toads overwinter in old upturned flowerpots, in piles of leaf litter or under large stones.

The first frosts are a signal to many that the time for preparation is past, and the moment has come for them to tuck themselves away. Everything either dies with the coming of the frosts or takes to the nooks and crannies around the pond, and the surface, once alive with pond skaters and whirligig beetles, is still and quiet. The pace of life has slowed dramatically, and the pond is ready for winter.

Great
reedmace

Greater
pond
sedge

Soft
rush

O

November

1 Samhain – celebration of the beginning of winter (Gaelic/Pagan)

1 All Saints' Day (Christian)

2 All Souls' Day (Christian)

4 Bridgwater Carnival

5 Guy Fawkes Night

11 Martinmas (Christian/traditional)

11 Remembrance Day

12 Remembrance Sunday

13 Diwali – festival of lights (Hindu/Sikh/Jain)

16 Beaujolais Nouveau Day

23 Thanksgiving (US traditional)

26 Stir-up Sunday

30 St Andrew's Day – patron saint of Scotland – bank holiday, Scotland

NOVEMBER AT A GLANCE

Historically this month has begun with a string of sombre celebrations. Allhallowtide encompasses All Saints' Eve (Hallowe'en) on 31st October, All Saints' Day on 1st November and All Souls' Day on 2nd November; the older, Gaelic festival of Samhain (pronounced sah-win) ran through the same dates. What all of these festivals have in common is an interest in the otherworldly and the departed. Long ago, particularly in Ireland and on the western fringes of Britain, Samhain was one of the biggest festivals of the year: a great fire celebration to mark the beginning of winter. It was considered a magical, liminal time, as the changeovers between the seasons often were. This was a time to mark and mourn the dead, but also to beware of evil spirits and perhaps make offerings to the fairies – the *aos sí* – or little folk.

This preoccupation with death echoes what is happening in the countryside around us this month. The final leaves are tumbling from the trees and very soon all that will be left is the barest countryside, brown and black and apparently dead. It's not surprising that an element of death anxiety filtered into this month's seasonal celebrations.

And so perhaps this is a natural moment for us, too, to contemplate loss. We are all saying goodbye to the summer, to its ease and light and good things, and entering into harder, colder days. But if that has echoes for you and your life, then try to take comfort from the year, and from the knowledge that – although things may be hard now and in your immediate future – gentler, kinder times will come again. They always do. For now, light a fire or a candle and spend some time with the thoughts of what you have to say goodbye to.

THE SKY AT NIGHT

Jupiter will be at opposition (see page 172) early this month, and so is at its highest and brightest of the year.

3rd: Jupiter at opposition: close, high, bright and visible all night. This is the time to look at it through binoculars or a telescope, when you may even pick out its largest moons.

4th–5th: Taurids meteor shower. Running from 7th September to 10th December, this is a minor meteor shower that produces up to ten meteors at its peak on the night of the 4th and early hours of the 5th November. There will be a lot of light from the moon so it may be tricky to spot.

9th: Close approach of the moon and Venus. They rise at around 03.00 in the east, then become lost in the dawn at about 06.40 in the southeast at an altitude of 34 degrees above the horizon. There is a lunar occultation of Venus (in which the moon passes in front of Venus) in the morning, but it will be tricky to see as the sun will be at its highest.

13th: Uranus at opposition. It will be brighter than at any other time of year, but due to its vast distance away from earth it will only be visible with a powerful telescope under very good dark-sky conditions.

17th–18th: Leonids meteor shower. Up to 15 meteors per hour can be produced by this meteor shower. Running from 6th to 30th November, it peaks on the night of the 17th and early morning of the 18th. Every 33 years it produces a particularly spectacular display, but the last one happened in 2001 so we have a bit of a wait yet.

20th: Close approach of the moon and Saturn. They first appear in the dusk at around 16.30 in the southeast at an altitude of 19 degrees above the horizon. They reach their greatest altitude of 24 degrees at about 18:20 in the south, then set at about 22.40 in the southwest.

25th: Close approach of the moon and Jupiter. They first appear in the dusk at around 16.30 in the east at an altitude of 14 degrees above the horizon. They reach their greatest altitude of 54 degrees at about 22.20 in the south. They set at about 04.50 the next day in the west.

N

THE SOLAR SYSTEM

Jupiter

Vast and beautiful Jupiter is not only the largest planet in the solar system – at two and a half times the mass of all the other solar system objects combined apart from the sun – but is also thought to be the oldest. As such, it has had an impact on the development of all the other planets, not least earth.

When the solar system was young, a growing Jupiter began a great migration towards the sun, the first stage in what is known as 'the grand tack'. Sweeping material up as it went, it deprived the other growing planets of material – which may be the reason Mars is so small. But as it did this, Saturn was also growing, and eventually its gravitational pull helped to haul Jupiter back out towards the outer solar system. These two great disturbances scattered the asteroid belts, and forced icy asteroids from the outer solar system, where ice was abundant, inwards towards the rocky planets, where it was not. They may have been responsible for first delivering water to earth.

Jupiter is at opposition early this month and will be high and bright in our skies all night long.

Sunrise and set
Haltwhistle, Northumberland

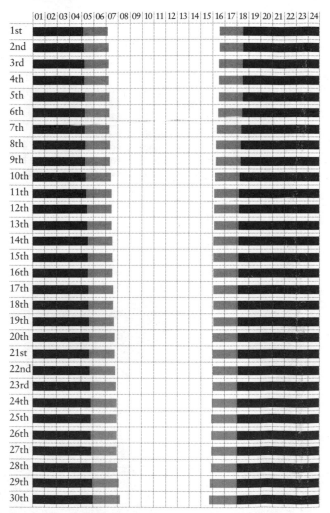

THE SEA

Average sea temperature in Celcius

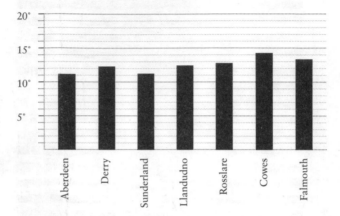

Spring and neap tides

Spring tides are the most extreme tides of the month, with the highest rises and the lowest falls, and they follow a couple of days after the full moon and new moon. These are the times to choose a low tide and go rock-pooling, mudlarking or coastal fossil-hunting. Neap tides are the least extreme, with the smallest movement, and they fall in between the spring tides.

Spring tides: 14th–15th and 28th–29th
Neap tides: 6th–7th and 21st–22nd

Spring tides are shaded in black in the chart opposite.

November tide timetable for Dover

For guidance on how to convert this for your local area, see page 8.

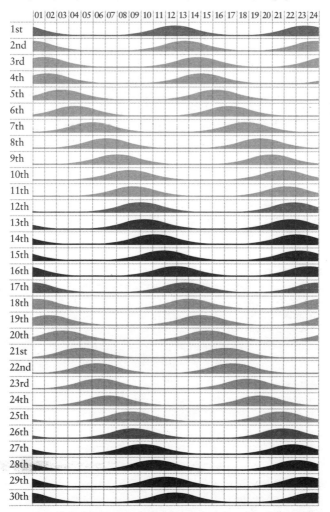

THE MOON

Moon phases

Last quarter – 5th November, 08.37

New moon – 13th November, 09.27

First quarter – 20th November, 10.50

Full moon – 27th November 09.16

Moonrise and set

Like the sun, the moon rises roughly in the east and sets roughly in the west. It also rises around 50 minutes later each day. Use the following guide to work out approximate moonrise times.

Full moon: Rises near sunset, opposite the sun, so in the east as the sun sets in the west.
Last quarter: Rises around midnight, and is at its highest point as the sun rises.
New moon: Rises at sunrise, in the same part of the sky as the sun (and so cannot be seen).
First quarter: Rises at around noon, and is at its highest point as the sun sets.

Full moon

November's full moon is known as the Darkest Depths Moon or Mourning Moon. This one is also the Moon Before Yule.

New moon

This month's new moon, on the 13th, is in Scorpio. Astrologers believe that the new moon is a quiet, contemplative time before a phase of growth. Each new moon has its own energy, depending on the zodiacal sign that it is in, and the Scorpio new moon is said to rule intensity and sensuality.

Moon phases for November

1st	2nd	3rd	4th	5th
6th	7th	8th	9th	10th
11th	12th	13th NEW	14th	15th
16th	17th	18th	19th	20th
21st	22nd	23rd	24th	25th
26th	27th FULL	28th	29th	30th

N

GARDENS

To enjoy this month

Ornamental: Chrysanthemums, rosehips, crab apples, spindle, old man's beard, winter pansies and violas, dogwood stems, final autumn leaves, ornamental grasses, teasel, lords and ladies, holly, ivy
Edible: Kale, cabbages, Brussels sprouts, carrots, beetroot, maincrop potatoes, Oriental leaves, pumpkins, winter squashes, celeriac, scorzonera, salsify, ceps, chanterelles, puffballs

Gardening by the moon
The following is a guide to planting with the phases of the moon, according to traditional practices. It also works as a guide to the month's gardening for moon-gardening cynics, who can do these jobs whenever they wish during the month ahead.

Full moon to last quarter: 29th October–4th and 27th–4th December
The waning moon is traditionally associated with a 'drawing down' energy, and this phase is a good time for sowing and planting any crops that develop below ground: root crops, bulbs and perennials. Light is high but decreasing.
- Plant tulips, lily and ornamental allium bulbs.
- Plant up paperwhite, miniature iris and forced hyacinth bulbs for late-winter indoor flowers.
- Plant garlic cloves and overwintering onion sets.
- Plant rhubarb crowns and bare-root fruit bushes.
- Plant new grapevines, peaches, nectarines.
- Plant apples, pears, quinces and medlars.
- Plant new flowering perennials. Lift, divide and replant those that have finished flowering.

Last quarter to new moon: 5th–12th
A dormant period, with low sap and poor growth. Do not sow or plant. A good time though for pruning, while sap is slowed. Weeding now will check growth well. Harvest any crops for

storage. Fertilise and mulch the soil. Garden maintenance.
- Prune apple, pear, medlar and quince trees.
- Prune grapevines before the winter solstice.
- Remove the nets from fruit cages. Prune autumn-fruited raspberries, red and white currants, and gooseberries.
- Remove unripened figs.
- Check your soil for its pH level. If it is low add lime or calcified seaweed.
- Mulch beds with organic manure.
- Pick apples and pears for storage.
- Lift dahlias once blackened by first frosts. Dry out and store the tubers in moist sand somewhere cool but frost-free.
- Prune roses.
- Collect fallen leaves into leaf mould bins or bin bags.

New moon to first quarter: 13th–20th (till 10.50)
The waxing of the moon is traditionally associated with rising vitality and upward growth, and towards the end of this phase is a good time for planting and sowing anything that develops above ground. Prepare for growth.
- You can sow, plant out or take cuttings of all of those things mentioned in the first quarter to full moon phase, but towards the end of this phase is better.

First quarter to full moon: 20th (from 10.50)–26th
This is the best time for sowing crops that develop above ground, but is bad for root crops. Plant out seedlings and young plants. Take cuttings and make grafts. Avoid any other pruning. Fertilise.
- Sow broad beans for early summer crops next year, either direct into the ground or in pots in the greenhouse.
- Sow sweet peas and hardy peas in the greenhouse.

Note: Where no specific time for the change between phases is mentioned, this is because it happens outside of sensible gardening hours. For exact changeover times, refer to the moon phase chart on page 243.

N

THE RECIPES

Bun of the month

Pan de Muertos – Day of the Dead buns

These buns flavoured with orange, star anise and honey are eaten in Mexico on the Day of the Dead (*Día de Muertos*), which actually encompasses the three days of Allhallowtide: All Saints' Eve (Hallowe'en), All Saints' Day and All Souls' Day: 31st October–2nd November. In Mexico these days are a colourful and lively celebration of the dead, with families visiting the graves of loved ones and telling happy stories about them. *Pan de Muertos*, often decorated with pieces of dough shaped to look like bones, is given to loved ones and eaten dipped in coffee or hot chocolate. It is traditionally made with anise, rather than star anise, but anise is difficult to get hold of. If you have it, then substitute 1 tablespoon of ground anise for the star anise.

Makes 12
Ingredients
60ml water
160ml rapeseed oil
100g clear honey
1 star anise
Zest and juice of 1 orange
1 teaspoon orange extract
400g strong white bread flour, plus extra for dusting
1 sachet (7g) instant yeast
1 teaspoon salt
5 eggs
A splash of milk
50g butter, melted
Icing sugar, for sprinkling

Method

Line a baking tray with parchment. In a saucepan, gently heat the measured water, oil, honey and star anise until the honey has dissolved. Add the orange zest and juice and the orange extract, and leave to cool and infuse.

Combine the flour, yeast and salt in a bowl. Make a well in the middle of the flour mixture, and crack in 4 eggs. Remove the star anise and pour in the cooled liquid, combining all into a wet dough. You want to keep it as pliable as possible, so don't add too much flour – just enough to be able to handle it. Turn onto a well-floured surface and gently knead for a few minutes, then pop it back in the bowl, cover and leave in a warm place for an hour.

Form the dough into 13 balls, one of which is double the size of the others and will be used to decorate the top of the buns. Smooth the 12 balls and place on the baking tray, leaving room for them to grow, then brush them with egg wash (1 egg beaten with a splash of milk). From the large remaining ball, make 12 small balls, and also roll out 24 dough 'worms'. Place two 'worms' in a cross over the top of each bun. Now stick one of the 12 small balls in the middle of each of the crosses. These are meant to look like bones, so squeeze them into different shapes to resemble bony fingers. Allow them to puff up again for an hour.

Meanwhile, preheat the oven to 200°C, Gas Mark 6. When the buns have risen, bake them for about 15 minutes, then leave to cool. Brush with melted butter, sprinkle with icing sugar, then serve.

N

Walnut and Parmesan *gougères* for *Beaujolais Nouveau* Day

On the third Thursday in November, the year's *Beaujolais nouveau* – the first wine of the season – is released. *Beaujolais nouveau* is a light, cherry-coloured wine made to be drunk young; the grapes will have been harvested just months before. Such *vin de primeur* wines, intended to be drunk within six months of harvesting, would originally have been imbibed by the vineyard workers to celebrate the end of the harvest, and so are the perfect focal point for a little party. You might want to offer some French charcuterie, pâtés, cheese, cornichons and olives, and these walnut and Parmesan *gougères*, which, like Beaujolais, originate in Burgundy.

Makes 12–15
Ingredients
40g butter, cubed
1 teaspoon salt
125ml water
70g plain flour
2 large eggs, beaten
½ teaspoon cayenne pepper
40g walnuts, finely chopped
70g Parmesan cheese, finely grated

Method

Line a large baking tray with parchment and preheat the oven to 220°C, Gas Mark 7.

Put the butter, salt and measured water in a small saucepan over a low heat, stirring occasionally until melted. Remove from the the heat and add the flour, stirring for a minute or so until it forms a paste that lifts cleanly off the sides of the pan. Use an electric whisk to beat the mixture for a minute, then add the eggs, a little at a time, until you have a pliable mix.

Fold in the cayenne, the walnuts and the bulk of the cheese. Using a damp tablespoon, scoop small dollops onto the baking tray, giving each one space to puff and expand. (Alternatively, you can pipe them onto the tray if you'd prefer them to be appear more uniform in size.) Sprinkle over the last few bits of cheese. Turn the oven down to 200°C, Gas Mark 6, and bake near the top of the oven for 25 minutes.

Remove the buns, carefully pricking each one with a sharp knife to let any steam escape and to reduce the chances of sogginess. Transfer to a cooling rack before serving.

THE ZODIAC

Scorpio: 23rd October–21st November

The sun begins the month in the same area of sky that holds the constellation of Scorpio, the Scorpion, the 210th–240th degree of the zodiac.On the 22nd of this month the sun will move into Sagittarius (see page 275).

> **Symbol:** The Scorpion
> **Planet:** Pluto, Mars
> **Element:** Water
> **Colour:** Black, deep red
> **Characteristics:** Passionate, intense, extreme, emotional, uncompromising, funny, sexy

Orion – the son of the Greek sea god, Poseidon, and the mortal Euryale – was a great hunter, and a favourite companion of Artemis, the Greek goddess of the hunt. They were hunting together on the island of Crete when Orion boasted that he could kill every creature on earth. Gaia, the earth mother, was naturally not too pleased to hear this, and so she conjured up a great scorpion to battle Orion and stop him. The scorpion killed him, and Zeus placed both Orion and the scorpion in the sky as the constellations Orion and Scorpius – at opposite sides of the sky, to keep them apart.

The best time to spot Scorpio is when it is in the opposite part of the sky from the sun six months from now, in May. You could look now for May's zodiacal constellation, Taurus. This is also a great time to see Orion, which is one of the best winter constellations.

A FOLK SONG FOR SCORPIO'S SCORPION

'Hark, the Goddess Diana'
Traditional, arr. Richard Barnard

Unsurprisingly, there are no British or Irish folk songs that mention scorpions, and so here is a song to commemorate the great hunt of Orion and Artemis, before Gaia was moved to intervene with the scorpion in the tale behind the zodiacal sign of Scorpio. Artemis' Roman equivalent was Diana, and here she is invoked as the spirit of the hunt.

Hark, the Goddess Diana calls out for the chase,
Bright Phoebe awakens the morn.
Rouse, rouse from your slumbers to hunting prepare,
For the huntsman is winding his horn.

See the hounds are unkennelled and ready for chase,
They start to o'ertake the fleet hare.
All danger they're scorning in the hunt this fair morning,
To the fields then away let's repair.

N

NATURE

The pond in November

The leaves are falling fast now as the temperature and light levels drop. The water of the pond is starting to cool dramatically, and to survive it the pond and the creatures in and around it are pretty close to completely dormant. This is where the preparations you have made for wildlife in the garden really come into their own. All summer long the creatures of the pond and the garden have been happily living their lives, feeding and breeding, but now the edges of the garden become their refuge from the cold, and the messier you have left them, the better. It is not only amphibians and mammals that make use of these nooks and crannies. Pond skaters and other aquatic invertebrates overwinter, and will often cluster themselves into a nook in a woodpile. Leaf piles, compost heaps and upturned broken pots all have their parts to play, and if you have not put any into place it isn't too late to knock some together now. A great number of creatures also see out the winter in the muddy bottom of the pond, popping up to the surface every now and then to take a breath of air, and then sinking back into the depths. The frogs that have hunkered down at the bottom will move around on sunny and warm days. Look out for one swimming across the bottom when the weather is mild, or popping up to the surface for a breath of air.

Foxes, which do not hibernate, make use of the pond now, making night-time visits to its edges to take a drink, and you may notice their footprints in the mornings in the mud surrounding the pond. You may also see the footprints of blackbirds around the pond edges, where they will pick through the mud and plants searching for grubs and insects.

Grey heron

Urban fox

Grey squirrel

Hedgehog

N

December

 Start of meteorological winter

 First Sunday in Advent (Christian)

 Hanukkah – festival of lights (Jewish) begins at sundown

22 Winter solstice, 03.27

24 Christmas Eve (Christian)

25 Christmas Day (Christian) – bank holiday England, Scotland, Wales, Northern Ireland, Ireland

26 St Stephen's Day/Boxing Day (Christian) – bank holiday England, Scotland, Wales, Northern Ireland, Ireland

31 New Year's Eve

DECEMBER AT A GLANCE

There is a great contradiction to December. It is, of course, our darkest, gloomiest month, and if you were a caveman matching your energy to the season you would think that this would be a time to retreat to your cave and wait out the darkness until spring. But December doesn't roll like that, and it hasn't done for many hundreds of years; there has to be something in it. There is, of course, the little matter of Christmas at the end of the month, but people have been gathering together and celebrating this darkest moment in the year long before Christianity came along. There appears to be something about reaching the deepest depths of the year that inspires us to to use up an over-generous proportion of our precious resources in gathering, feasting and celebrating. And so this month we come to glittering, sparkling life, just as nature reaches its most complete hibernation. We light up our homes with twinkling lights, candles and log fires, and invite people in. We fight the dark. Later on this month we will get a chance to play at hibernation, possibly our best chance in the year, in that glorious week of nothing between Christmas and New Year, when – if you are very lucky – the only things on the agenda are an old film, a walk, a fire and a small mountain of leftovers. But historically this month has been about celebrating our having reached the nadir, and revelling in the fact that from here onwards our world will get incrementally lighter and brighter, and remarkably quickly, too. At midwinter we have a half a year of lengthening days ahead of us, and that is a joyous thought. If lighting up the darkness to mark this moment wasn't in some way necessary to us, we wouldn't have been doing it for so very, very long.

THE SKY AT NIGHT

There are opportunities to see several of the bright planets in our skies this month, but the highlight is the Geminids, a reliable and spectacular meteor shower that falls on a night with very little moon: the perfect set-up for a fantastic show, if we get some clear weather.

4th: Mercury at greatest eastern elongation. Mercury is the closest planet to the sun, and this makes it hard to spot, as it is usually lost in the sun's glare. When it is at its furthest point from the sun in our sky, there is a chance to see it for a week or so. Spot it low in the western sky shortly after sunset.

9th: Close approach of the moon and Venus. They rise at around 04.20 in the east and become lost in the dawn at about 07.20 in the southeast at an altitude of 23 degrees above the horizon.

13th–14th: Geminids meteor shower. One of the best meteor showers of the year, it produces up to 120 multicoloured meteors at its peak, created when the earth moves through the dust trail left by asteroid 3200 Phaeton. The dust burns up as it hits the atmosphere. It runs from around 7th to 17th December, peaking on the late night and early morning of the 13th and 14th. This will be just past the new moon so there will be no interference from moonlight, potentially making for a spectacular show.

17th: Close approach of the moon and a dim Saturn. They first appear in the dusk at around 16.20 in the south at an altitude of 23 degrees above the horizon. They set at about 21.00 in the southwest.

22nd: Close approach of the moon and Jupiter. They first appear in the dusk at around 16.20 in the east at an altitude of 30 degrees above the horizon. They reach their greatest altitude of 54 degrees at about 20.20 in the south. They set at about 03.00 the next day in the west.

D

THE WINTER SOLSTICE

Again we reach the point this month when the axial tilt of the earth is aligned with the sun, only this time the north pole is tipped away from it and the south pole is tipped towards it. Historically this moment has been about gathering and feasting, and lighting fires to celebrate the coming of the light. The 'circle of latitude' that is of most importance to the winter or December solstice is the Tropic of Capricorn, which is 23.43 degrees south of the equator and runs through southern Africa, Madagascar, the Indian Ocean, Australia, the Pacific Ocean and South America. This is the most southerly point on earth where the sun can be directly overhead, and it does this only at the exact moment of the winter solstice, before heading back north again.

On the solstice, any point along the Antarctic Circle will have 24 hours of light, and just like the midnight sun in the Arctic Circle in summer, the further south you go into the Antarctic, the more days and nights of 24-hour light there are – this phenomenon is less known simply because there are no permanent communities within the Antarctic Circle, and the only human habitation is in research stations, unlike within the Arctic Circle, which contains a number of towns and cities. On the date of the winter solstice, the Arctic Circle and all within it will experience 24 hours of darkness.

Winter solstice facts
Date: 22nd December
Time: 03.27
Altitude: At solar noon on 22nd December, the sun will reach 15 degrees in the London sky, 11 degrees in the Glasgow sky and 13 degrees in the Dublin sky.
Sunrise times: London 08.04, Glasgow 08.46, Dublin 08.38
Sunset times: London 15.53, Glasgow 15.44, Dublin 16.08
Sun's distance from earth: 147,160,000 kilometres

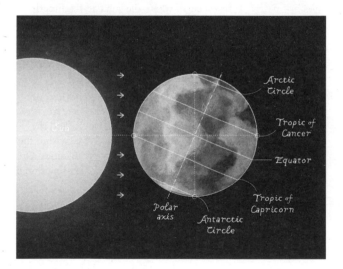

Mark the winter solstice

- Make a nature table with pieces of lichen-covered moss, berried holly, ivy, seed heads, pebbles, old man's beard, feathers and whatever else you can find out-of-doors, then place white candles among them and light them on the day.
- It is wonderful to catch the sunrise on the winter solstice, even if only from your window, so do it if you can – it certainly happens late enough. As you greet the sun, think about how it will make its lowest arc in the sky today, but how soon it will start to rise higher, lengthen our days and bring spring.
- Fill a spray bottle with water, 1 teaspoon of vodka and 25–30 drops of Christmassy essential oils, then shake it up and spritz it around. Try: frankincense, sweet orange, Douglas fir.

Sunrise and set
Haltwhistle, Northumberland

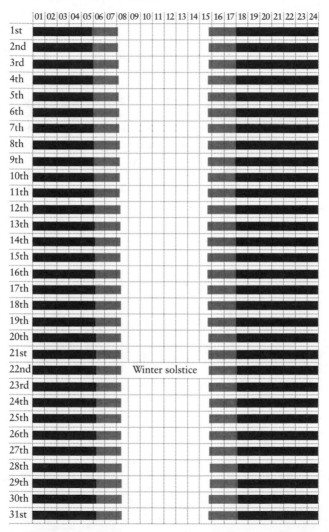

	01	02	03	04	05	06	07	08	09	10	11	12	13	14	15	16	17	18	19	20	21	22	23	24
1st																								
2nd																								
3rd																								
4th																								
5th																								
6th																								
7th																								
8th																								
9th																								
10th																								
11th																								
12th																								
13th																								
14th																								
15th																								
16th																								
17th																								
18th																								
19th																								
20th																								
21st																								
22nd						Winter solstice																		
23rd																								
24th																								
25th																								
26th																								
27th																								
28th																								
29th																								
30th																								
31st																								

D

THE SEA

Average sea temperature in Celcius

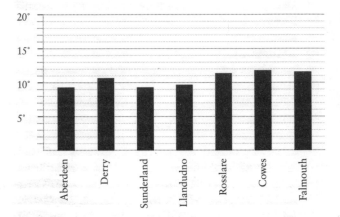

Spring and neap tides
Spring tides are the most extreme tides of the month, with
the highest rises and the lowest falls, and they follow a couple
of days after the full moon and new moon. These are the
times to choose a low tide and go rock-pooling, mudlarking
or coastal fossil-hunting. Neap tides are the least extreme,
with the smallest movement, and they fall in between the
spring tides.

Spring tides: 13th–14th and 28th–29th

Neap tides: 6th–7th and 20th–21st

Spring tides are shaded in black in the chart opposite.

December tide timetable for Dover

For guidance on how to convert this for your local area, see page 8.

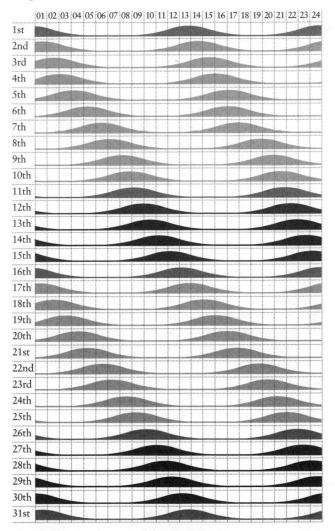

THE MOON

Moon phases

Last quarter – 5th December, 05.49

New moon – 12th December, 23.32

First quarter – 19th December, 18.39

Full moon – 27th December, 00.33*

Moonrise and set

Like the sun, the moon rises roughly in the east and sets roughly in the west. It also rises around 50 minutes later each day. Use the following guide to work out approximate moonrise times.

Full moon: Rises near sunset, opposite the sun, so in the east as the sun sets in the west.
Last quarter: Rises around midnight, and is at its highest point as the sun rises.
New moon: Rises at sunrise, in the same part of the sky as the sun (and so cannot be seen).
First quarter: Rises near noon, and is at its highest point as the sun sets.

Full moon

December's full moon is known as the Oak Moon or Full Cold Moon. This one is also the Moon After Yule.

New moon

This month's new moon, on the 12th, is in Sagittarius. Astrologers believe that the new moon is a quiet, contemplative time before a phase of growth. Each new moon has its own energy, depending on the zodiacal sign that it is in, and the Sagittarius' new moon is said to rule optimism and trust.

Moon phases for December

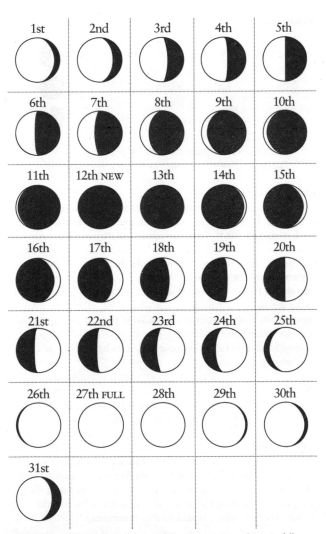

*This full moon falls in the early hours of the morning. To catch it at its fullest during normal waking hours, view it the evening before.

GARDENS

To enjoy this month

Ornamental: Holly, mistletoe, ivy and its flowers and berries, seed heads, willow stems, dogwood stems, rosehips, haws, old man's beard, winter clematis, mahonia, winter jasmine, chimomanthus, sarcococca, box, witch hazel, floral plumes of ornamental grasses, skeletons of trees, crab apples, cotoneaster, pyracantha
Edible: Quinces, beetroot, Brussels sprouts, maincrop potatoes, carrots, kale, chard, garlic, leeks, parsnips, swedes, pumpkins, winter squashes, chervil, parsley, coriander, sage, rosemary, bay

Gardening by the moon
The following is a guide to planting with the phases of the moon, according to traditional practices. It also works as a guide to the month's gardening for moon-gardening cynics, who can do these jobs whenever they wish during the month ahead.

Full moon to last quarter: 27th November–4th and 27th–3rd January 2024
The waning moon is traditionally associated with a 'drawing down' energy, and this phase is a good time for sowing and planting any crops that develop below ground: root crops, bulbs and perennials. Light is high but decreasing.
- Plant garlic cloves.
- If the ground isn't frozen, plant new fruit bushes and trees.
- Plant rhubarb crowns. Lift, split and replant large clumps.
- Plant new perennials in your flower borders, and lift, divide and replant those that have finished flowering.

Last quarter to new moon: 5th–12th
A dormant period, with low sap and poor growth. Do not sow or plant. A good time though for pruning, while sap is slowed. Weeding now will check growth well. Harvest any crops for storage. Fertilise and mulch the soil. Garden maintenance.
- Prune apple, pear, medlar and quince trees.

- Prune grapevines before the winter solstice. The sap starts to rise very early in the new year and so they will 'bleed' if left any later.
- Check your soil for its pH level. If it is low, this would be a good time to add lime or calcified seaweed. This raises the pH, which in turn makes nutrients more easily available to plants. Brassicas in particular appreciate lime, but few vegetables grow well with a low pH.
- Mulch beds with organic manure.
- Weed thoroughly.
- Prune roses.
- Put terracotta pots onto pot feet to lift them from the ground and improve drainage, ahead of the very cold months ahead.
- Order seed for next year.

New moon to first quarter: 13th–19th
The waxing of the moon is associated with rising vitality and upward growth. Towards the end of this phase plant and sow anything that develops above ground. Prepare for growth.
- Sow microgreens on a windowsill: basil, dill, celery, onion, chervil, beetroot, coriander, red mustard and pea. Harvest when 5cm tall.

First quarter to full moon: 20th–26th
This is the best time for sowing crops that develop above ground, but bad for root crops. Plant out seedlings and young plants. Take cuttings and make grafts. Avoid any other pruning. Fertilise.
- Take grafts of favourite apple varieties to make new young plants.

Note: Where no specific time for the change between phases is mentioned, this is because it happens outside of sensible gardening hours. For exact changeover times, refer to the moon phase chart on page 267.

D

THE RECIPES

Bun of the month

Maple and saffron St Lucia buns

In Sweden and other Scandinavian countries, 13th December is St Lucia's Day. Children dress up in long white gowns, the girls with red ribbons tied at the waist, the boys wearing conical hats covered in gold stars. All carry white candles, and one girl wears a crown of white candles on her head and carries a tray of saffron buns. St Lucia was an early Christian martyr from Sicily whose eyes were put out when she refused to marry, choosing to dedicate her life to Jesus. After her death her eyes were miraculously restored. Quite how she ended up being so important in Scandinavia is a bit of a mystery, but she is associated with bringing light in the darkness, something that northern Europe is rather in need of at this time of year. Right up until 1753, when Sweden changed from using the Julian calendar to the Gregorian calendar, 13th December was midwinter, and this festival of candles would have brought light to the shortest day. It is considered the beginning of the Christmas season in Sweden.

Makes 12

Ingredients

250ml milk, plus extra for the glaze

50g maple syrup, plus extra for the glaze

Pinch of saffron

50g butter

500g strong white bread flour, plus extra for dusting

1 teaspoon salt

1 sachet (7g) instant yeast

100g natural yoghurt

Handful of sultanas, to decorate

Poppy seeds, for sprinkling

Method

Line a baking tray with parchment. Gently heat the milk and maple syrup with a pinch of saffron until steamy, then remove and add the butter. Allow to cool while you weigh out the flour, salt and yeast into a large bowl. Add the milk mixture and yoghurt to the flour, and combine to form a soft dough. Turn out onto a floury surface and knead for 5 minutes until the dough is smooth and elastic. Pop the dough back into the bowl, cover and set in a warm place for 2 hours.

Now knock the air out of the dough and divide into 12 pieces, rolling each into a 30cm-long sausage. Leave to rest for 10 minutes, then form your dough sausages into tight 'S' shapes on the tray. Cover and leave for a further 45 minutes to plump up again.

Meanwhile, preheat the oven to 200°C, Gas Mark 6. Make a glaze with a splash of milk and a little glug of maple syrup, and brush this over the buns. Nestle a sultana in each nook of the 'S' shapes and sprinkle with poppy seeds. Bake in the centre of the oven for 15 minutes or until golden on top. Allow to cool before serving.

D

Cumberland rum nicky

This pie originated in the 19th century in Cumberland, where it is often eaten around Christmas time. It is filled with dried tropical fruits, ginger, rum and sugar. These ingredients were rare and expensive in most of the country but came into Whitehaven, west Cumbria, because of the port's involvement in the 'triangular trading system' – the UK's routes for transporting enslaved people and goods between Africa, England and the Caribbean. The ingredients are therefore common in Cumbrian recipes. This recipe has been supplied by Anti Racist Cumbria (antiracistcumbria. org). The organisation prompts people to think about what we consider to be traditionally 'Cumbrian', and to understand that Cumbrian history is Black history, too.

Serves 6

Ingredients

For the filling:

225g dates, coarsely chopped

100g dried apricots, coarsely chopped

50g stem ginger in syrup, drained and finely chopped

50ml dark rum

50g soft dark brown sugar

50g unsalted butter, cut into 2-cm cubes

For the sweet shortcrust pastry:

200g plain flour, plus extra for dusting

2 tablespoons icing sugar

100g unsalted butter, chilled and cut into 1-cm cubes

1 egg, lightly beaten

1 teaspoon lemon juice

2 tablespoons cold water

For the rum butter:

100g unsalted butter, softened
225g soft light brown sugar
75ml dark rum

Method

Mix all the filling ingredients except the butter together in a
bowl and leave to soak. Meanwhile, make the pastry. Mix the
flour and icing sugar together in a bowl, add the butter and
rub it in between your fingertips until the mixture resembles
fine breadcrumbs.

In a separate bowl, mix the egg with the lemon juice and
the measured water. Make a well in the centre of the flour
mixture and pour in the egg mixture. Using a table knife,
work the liquid into the flour to bring the pastry together. Add
a splash more water if dry. When the dough begins to stick
together, use your hands to knead it into a ball. Wrap it in
clingfilm and rest in the refrigerator for at least 15 minutes.

Preheat the oven to 180°C, Gas Mark 4. Unwrap the dough,
cut it into 2 pieces: roughly one-third and two-thirds. Roll out
the larger piece on a floured surface and use it to line a 20-cm
pie dish. Leave any excess pastry hanging over the edge. Spread
the filling in the pastry and dot with the butter.

Roll out the remaining pastry and cut it into 8 long strips,
roughly 1cm wide. On a sheet of baking parchment, create
a lattice with 4 pastry strips going each way, weaving them
under and over each other. Dampen the edge of the pastry in
the tin, then invert the lattice from the paper onto the tart.
Press the ends of the strips to the pastry base to secure. Bake
for 15 minutes, then turn the oven down to 160°C, Gas Mark
3 and bake for a further 20 minutes.

Meanwhile, for the rum butter, beat together the butter
and sugar, before slowly adding the rum. Refrigerate until
needed. Serve the tart warm or cold, with the rum butter.

Clementine, nutmeg and bay leaf eggnog

Creamy, spice-infused eggnog started life as a drink for the British upper classes. In the 18th century it crossed the Atlantic to North America, where over time it became popular as a Christmas drink and is now sold in cartons in vast amounts every Christmas. Having faded out of popularity in the UK and Ireland, eggnog has never quite caught on again here, but it is very much worth revisiting. If you have guests coming, you can make this ahead of time and store it in the refrigerator.

Serves 8

Ingredients

100g soft brown sugar

2 eggs, separated

Zest of 2 clementines

1 bay leaf

Nutmeg

600ml whole milk

100 ml rum

100ml whipping cream

Method

In a large bowl, use an electric whisk to combine the sugar, egg yolks, zest, bay leaf and about a quarter of a nutmeg, finely grated. Whisk until creamy. Heat the milk in a saucepan until it almost comes to the boil, then pour it over the egg mixture, whisking all the time. Now pour it all back into the pan, heating and stirring until it is slightly thickened. Allow to cool completely, then add the rum and stir in. Beat the egg whites until they have formed soft peaks, and fold them into the mixture. Beat the cream until it has formed soft peaks and fold that in. When ready to serve, divide between 8 glasses and grate a little nutmeg on top.

THE ZODIAC

Sagittarius: 22nd November–21st December

The sun begins the month in the area of sky that holds the constellation of Sagittarius, the Archer, the 240th–270th degree of the zodiac. On the 22nd of this month the sun will move into Capricorn. Mercury is in retrograde from 13th December 2023 to 1st January 2024.

> **Symbol:** The Archer
> **Planet:** Jupiter
> **Element:** Fire
> **Colour:** Dark purple, dark blue
> **Characteristics:** Adventurous, independent, strong-willed, generous, truthful, energetic

The constellation of Sagittarius celebrates Chiron, the wise centaur who taught many Greek heroes, including Achilles, Patroclus, Odysseus, Ajax, Heracles, Theseus and Perseus. Chiron was injured when he dropped one of Heracles' arrows, which had been dipped into the Hydra's venom (see page 160), onto his foot. Despite being a great healer, he died after nine days, passing into the stars to become the constellation Sagittarius. The best time to spot Sagittarius is when it is in the opposite part of the sky from the sun six months from now, in June.

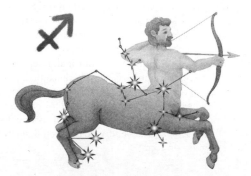

A FOLK SONG FOR
SAGITTARIUS' BOW

'Now Robin Lend to Me Thy Bow'
Traditional, arr. Richard Barnard

This song about one of the most famous archers of all – perhaps
on a par with Chiron, who is invoked in the story behind this
month's zodiacal sign, Sagittarius – was popular around the
time of Queen Elizabeth I. During this period Robin Hood was
a celebrated character, the subject of plays and songs, and even
receiving mentions in several of the works of Shakespeare. It can
be sung as a four-part round with voices entering after each line
of text.

'Now Robin lend to me thy bow,
Sweet Robin lend to me thy bow,
For I must now a-hunting go,
With my sweet Lady go.'

'And whither will thy Lady go?
Sweet Wilkie, tell it unto me.
You'll have my hawk, my hound, my bow
To wait on thy Lady.'

'My Lady will to Uppingham
To Uppingham I swear will she
And I am appointed as the man
To wait on my Lady.'

'Farewell, good Wilkie, on your way
For hunting never pleases me,
Beware thy hounds lead not astray
And anger thy Lady.'

'My hounds shall all be led in line
So well I can assure to thee,
Unless a fair pursuit I find
To please my sweet Lady.'

With that the Lady she came in
And willed them all for to agree
That hunting never was thought sin,
And never shall for me.

D

NATURE

The pond in December

The pond in the depths of winter would be a place of utter stillness, everything having sunk to the bottom or seeing the winter out in the pond's surroundings, if it weren't for the garden birds. Birds, of course, do not hibernate – they are active and highly visible in gardens all winter. A pond is a life-saver for winter birds. House sparrows will visit in noisy gangs, bathing in the shallows and drinking from the chilly water. Tits, too, gang together into groups of up to 20 in winter, and will make the most of the pond, as will goldfinches, feeding on the seed heads in the garden and then splashing and drinking from the pond. Blackbirds, robins and bramblings will visit regularly, as will migrants such as redwings and fieldfares. Moorhens may spend time in smaller ponds in the winter, as they tend to avoid larger bodies of water that become full of flocks of competing birds. Mallards also visit, but they can be destructive in a small pond. Birds turn this otherwise still and quiet period into a bit of a riot, from time to time at least.

One of the perils of really cold winter weather is that the pond can freeze over and leave birds no access to water, so during a freeze put out a container of water for them; change it if it, too, freezes, checking twice a day. This is not the only hazard of a frozen pond. In its depths are creatures in various states of suspended animation – frogs and newly mature newts, dragonfly nymphs, whirligig beetle eggs and more – and all of them need oxygenated water. A frozen surface reduces the amount of oxygen in the water and can also lead to a build-up of harmful gases from rotting vegetation. Extremely cold spells rarely last long, and it would be very unusual for the pond to be frozen over for the few weeks it would take for a dangerous build-up to occur. But if you are concerned, then make a hole in the ice by resting a pan of boiling water on it – not by cracking the ice, which can shock the sleeping inhabitants, down in the chill, murky depths.

Moorhen

Mallard

Coot

REFERENCES

Astronomical and calendarial information reproduced with
  permission from HM Nautical Almanac Office (HMNAO), UK
  Hydrographic Office (UKHO) and the Controller of Her
  Majesty's Stationery Office.

Moon and sun rises and sets and further calculations reproduced with
  permission from www.timeanddate.com.

Tidal predictions reproduced with permission from HMNAO, UKHO
  and the Controller of Her Majesty's Stationery Office.

Astronomical events are based on ephemerides obtained using the
  NASA JPL Horizons system.

Sea temperatures are reproduced with permission from
  www.seatemperatures.org.

ACKNOWLEDGEMENTS

Thanks to Richard Barnard, who approaches *The Almanac*'s folk songs each year with investigative rigour and musical knowledge, and has even written a couple of tunes for this edition, for those songs for which only the words were ever collected. Above and beyond, as ever!

It has been wonderful to work this year with Beth Al Rikabi on recipe development and testing. Thank you, Beth, for all of your hard work and creativity (and sorry about all the gluten!).

Thanks also to my dad, Jack Leendertz, who takes great care of the Sky at Night sections.

Thank you to Anti Racist Cumbria for the information and recipe for Cumberland rum nicky.

Thank you to Olia Hercules for her permission to adapt her *pyrizhky* (Ukrainian buns).

All of the above have been tweaked and edited by me and any mistakes are mine.

Whooli Chen's illustrations have delighted us all as they have come in. Magical, a little odd, perfect. Thank you, Whooli, for such wonderful work.

To the book's designer, Matt Cox of Newman+Eastwood, thank you for continuing to take such care over it.

To everyone at Octopus Publishing and Gaia, thank you so much for your diligent and enthusiastic shepherding of *The Almanac* out into the world each year: Stephanie Jackson, Jonathan Christie, Sarah Kyle, Kevin Hawkins, Matt Grindon, Karen Baker, Alison Wormleighton and Jane Birch. Thanks also to my agent, Adrian Sington at Kruger Cowne, for your guidance and support.

And finally, love and thanks to my family: Michael, Rowan and Meg, and to new additions Saffy and Korina, who forced me to take a great many more muddy walks during the writing of this almanac than I ever have during previous ones.

INDEX

ABOUT THE AUTHOR

Lia Leendertz is an award-winning garden and food writer based in Bristol. She presents a monthly podcast, 'As the Season Turns', about what to look out for in the month ahead. Her reinvention of the traditional rural almanac has become an annual must-have for readers eager to connect with the seasons, appreciate the outdoors and discover ways to mark and celebrate each month. Now established as the bestselling almanac in the market, this is the sixth instalment.

Find out more about Lia at:
www.lialeendertz.com
🐦 @lialeendertz
📷 @lia_leendertz

ABOUT THE ILLUSTRATOR

Whooli Chen is a Taiwan-based illustrator. She is a cat lover, a heavy reader and has an affection for plants and animals. Her art is soaked with surrealism, dream interpretation, botany, Asian traditional art and gentle colours.

Find out more about Whooli at:
www.centralillustration.com/illustration/whooli-chen
www.behance.net/whoolichen
📷 @whooli.chen

CALENDAR 2023

JULY

M	T	W	T	F	S	S
					1	2
3	4	5	6	7	8	9
10	11	12	13	14	15	16
17	18	19	20	21	22	23
24	25	26	27	28	29	30
31						

SEPTEMBER

M	T	W	T	F	S	S
				1	2	3
4	5	6	7	8	9	10
11	12	13	14	15	16	17
18	19	20	21	22	23	24
25	26	27	28	29	30	

NOVEMBER

M	T	W	T	F	S	S
		1	2	3	4	5
6	7	8	9	10	11	12
13	14	15	16	17	18	19
20	21	22	23	24	25	26
27	28	29	30			